Sustainable Agriculture, Food Security and Climate Change

Sustainable Agriculture, Food Security and Climate Change

– by –
Subhas Chand
Division of Soil Science

Lal Singh
Parmeet Singh
Division of Agronomy

Sher-e-Kashmir University of Agriculture Sciences and Technology of Kashmir
Shalimar Campus, Srinagar - 191 121
Jammu & Kashmir

2018
Daya Publishing House®
A Division of
Astral International Pvt. Ltd.
New Delhi - 110 002

First Published, 2012
Reprinted, 2018

ISBN: 9789351241942 (Hardbound)

Publisher's note:

Every possible effort has been made to ensure that the information contained in this book is accurate at the time of going to press, and the publisher and author cannot accept responsibility for any errors or omissions, however caused. No responsibility for loss or damage occasioned to any person acting, or refraining from action, as a result of the material in this publication can be accepted by the editor, the publisher or the author. The Publisher is not associated with any product or vendor mentioned in the book. The contents of this work are intended to further general scientific research, understanding and discussion only. Readers should consult with a specialist where appropriate.

Every effort has been made to trace the owners of copyright material used in this book, if any. The author and the publisher will be grateful for any omission brought to their notice for acknowledgement in the future editions of the book.

Published by : **Daya Publishing House®**
A Division of
Astral International Pvt. Ltd.
– ISO 9001:2015 Certified Company –
4736/23, Ansari Road, Darya Ganj
New Delhi-110 002
Ph. 011-43549197, 23278134
E-mail: info@astralint.com
Website: www.astralint.com

Digitally Printed at : **Replika Press Pvt. Ltd.**

Preface

Agricultural sector and allied activities contributing about 19 per cent in national economy and sustaining life of > 65 per cent of Indian farmers. The healthy and nutritious food is a primary right of every mankind besides livelihood security. Fostering the growth of national and global food supplies is necessary for eliminating hunger and reducing poverty, but it is not enough. Today, even in the midst of sufficient global food supplies, 800 million people are hungry because they cannot afford to buy the food for a healthy life. More than two billion people are at risk from micronutrient deficiencies (*e.g.* vitamin A, iodine, iron, zinc, manganese etc.), and more than 1 billion are actually disabled by mental retardation, learning problems and blindness. Indian farming community is facing the second generation problems like decreasing factor productivity, lowering soil fertility and soil biodiversity, saline and alkali soils, scarcity of irrigation water, nutrients removal and addition gap widen etc. Production based livelihood security is an important area having capability of providing ample employment opportunities for living a healthy life. The country like India and China have good manpower for production based systems. Agricultural production systems provide livelihood security without high risk of health at village level without any specific technical qualifications. The population growth has put tremendous pressure on the quality of Environment of urban life. The impact created by these wastes on the environment is enormous, if proper disposal and management options are not applied. The waste could become a resource of bulk organic nutrients and the society can benefit from these wastes with proper collection and disposal technologies. The book is a classical collection of information regarding various approaches which are currently in the operation for enhancing food and nutritional security. First three chapters are described about different types of food security and production system.

The population growth has put tremendous pressure on the quality of environment of urban life. The residents generate various kinds of wastes of biodegradable and non biodegradable categories. The chapter 4 is useful for readers interested in waste disposal and management for environmental security. Climate change is a reality world wide attracting attentions of various forms to overcome and mitigate its impact to make harmonious environment and cooling earth for well being of mankind, hence chapter 5 and 6 exclusively prepared for readers, who are interested in climate change in relation to agriculture technology. The precision farming for natural resource management and sustainable crop production has also been described and information, reviews are immensely useful for readers of various expertises. Resource conservation technologies are given in details to update recent knowledge regarding various conservation practices in chapter 8.

Application of all the needed nutrients through fertilizers had adverse effect on soil fertility leading to unsustainable production for longer time; while integration of chemical fertilizers with organic manures not only maintain soil fertility but sustain crop productivity also. An efficient nutrient management system would minimize loss of nutrients, saving unnecessary input cost. Efficient nutrient use is essentially an offspring of balanced nutrient use and sound management practices and decisions. Balanced nutrient use is not only the first requirement; it is rather a pre-requisite since no amount of agronomic manipulation can produce high efficiency out of an imbalanced nutrient dose. Crop productivity and soil fertility are interlinked. But ultimately it is improvement in the nutrient-use efficiency that will decide crop production, productivity per unit area and long term sustainability on long-term basis. Crop production involves a combination of practices revolving around soil, crop, climate and management factors. The factors affecting crop yields and environmental sensitivity vary in both space and time. Thus a new concept has arisen to manage the space-time continuum in crop production called precision agricultural management or precision agriculture or site specific management, which is concerned with the management of variability of agricultural resources in the dimensions of both space and time. The advances in nutrient management are given in chapter 9 and principle and practices of organic farming are given in chapter 10.

Authors acknowledge the writers of different resources of literatures which may have been consulted during the preparation of manuscript. The authors hopes that book shall be useful for agronomist, soil scientist, environmentalists, agrometeorologist, researchers, students, progressive farmers, teachers of agricultural colleges and universities, employees of agricultural departments, policy planners and knowledge seekers etc.

Subhash Chand

Lal Singh

Parmeet Singh

Contents

Abbreviations

ACPA: Australian center for precision agriculture
ADB: Asian development bank
AHAF: Agri-horticulture agro forestry system
AMFUs: Agromet field units
APEDA: Agriculture processed product export developing agency.
ATMA: Agricultural technology management agency
BCR: Benefit Cost Ratio
BGA: Blue green algae
BMPs: Best fertiliser management practices
BMSF: Biological management of soil fertility
BNF: Biological nitrogen fixation
CA: Conservation agriculture
CCUBGA: Center for conservation and utilisation of blue green algae
CDM: Clean development mechanism
CEC: Cation exchange capacity
CGIAR: Consultative group on international agriculture research

CIAE: Central institute of agriculture engineering
CIMOD: Center for Integrated Mountain Development
CNBFES: Carbon neutral bio-diesel fuelled energy system
CNG: Compressed natural gas
CRM: Crop residues management
CS: Carbon sequestration
CSA: Clay settling areas
CSSRI: Central soil salinity research institute
CT: Conventional tillage
DPL: Dual purpose legumes
DRIS: Diagnosis and recommended integrated system
DSR: Direct seeded rice
DSS: Decision support system
DWR: Directorate of wheat research
ECC: Enriched city compost
EFTOs: Eco-friendly tillage options
ESP: Exchangeable sodium percentage
FAO: Food and agriculture organization
FC: Farming carbon
FCI: Food corporation of India
FIRBS: Furrow irrigation raised bed planting system
FP: Farmer practices
FUEs: Fertilizer use efficiencies
FYM: Farm yard manure
GDP: Gross domestic product
GHGs: Green house gases
GIS: Geographical information system
GLONASS: Global navigation satellite system
GM: Green manuring/genetically modified
GPS: Geographical positioning system
GtC: Gega tera gram
IARI: Indian agriculture research institute
IBI: International biochar initiatives

ICRISAT: International crop research institute for semi arid tropics
ICSWEQ: International conference on soil water environmental quality
IFAD: International fund for agriculture research
IFPRI: International food policy research institute
IGP: Indo-gigantic plains
IISS: Indian institute of soil science
IMD: Indian meteorology department
IMF: International monitory fund
INM: Integrated nutrient management
IPCC: Inter-Governmental panel on climate change
IPM: Integrated pest management
IPNM: Integrated plant nutrient management
IPNS: Integrated plant nutrient supply system
IRFT: International research for fair trade
IRRI: International rice research institute
ISFM: Integrated soil fertility management
ISRO: Indian space research organizations
IWM: Industrial waste material
KVK: Krishi vigyan Kendra
LAR: Local adhoch recommendations
LCC: Leaf colour chart
LEISA: Low external input sustainable agriculture
LLL: Laser land leveling
LUMPs: Land use management practices
LUPs: Land use practices
MNCs: Multinational companies
MOEF: Ministry of environment and forestry
MSW: Municipal solid waste
MTI: Miscellaneous topic of interest
NAAS: National academy of agriculture science
NABARD: National bank for agricultural rural development
NATP: National agricultural technology project

NCMRWF: National center for medium range weather forecasting
NMPs: Nitrogen management practices
NRAA: National rainfed area authority
NSOP: National standards of organic products
NT: No-till
OFDS: Optimising fertilizer doses
PCS: Portable personnel computer
PF: Precision farming
PHT: Post harvest technology
PLMZ: Production level management zones
PME: Post methanation distillery effluent
PPM: Part per million
PPS: Precesion water management
PSMs: Phosphate solubilising microorganisms
PTO: Power takes off
PWM: Precision water management
RAPs: Recommended agricultural practices
RCM: Research council meeting
RCTs: Resource conservation technologies
RDD: Rotary disc drill
RMPs: Recommended management practices
RS: Remote sensing
RW: Rice-wheat
SCS: Soil carbon sequestration
SCSPs: Soil carbon sequestration practices
SH: Soil health
SOC: Soil organic carbon
SOM: Soil organic matter
SPS: Standard positioning services
SQ: soil quality
SRI: System of rice intensification
SSNM: Site-specific nutrient management

SSSA: Soil science society of America
SSWM: Site specific water management
STCR: Soil test crop response
STLs: Soil testing labs
STV: Soil test value
Tg C/yr: Terra gram carbon per year
TL: Traditional levelling
UN: United Nations
UNCCD: United Nations convention to combat desertification
UNMDG: United Nations millennium development goals
UVR: Ultra-violet radiations
VRT: Variable rate technology
WB: World Bank
WTO: World trade organization
ZT: Zero-Tillage

Chapter 1

Household Food, Nutritional and Livelihood Security through Agricultural Interventions

"Agriculture is the backbone of the Indian economy and the villages are the life lines of growth of India"

—*M.K. Gandhi*

World is organizing 16 October as World Food Day every year to discuss and creates awareness about food and nutritional security and finding out appropriate strategies for meeting the food need of the world human population. The celebration should be seen in the context of world food and nutritional security. I would like to mention here that in the scenario of climate change and resource degradation, the goal is difficult, but achievable. The unprecedented increase in population has arrested all hopes of availability of balance diets. Indian farming community is facing the second generation problems like decreasing factor productivity, lowering soil fertility and soil biodiversity, saline and alkali soils, scarcity of irrigation water, nutrients removal and addition gap widen etc. Indian agriculture is also facing the implications of WTO agreement and Kyoto protocol (Reducing emission Green House (GH) Gases and decreasing capital investment due to high risk in agriculture. Chapter deals some suggestive solutions for food, nutritional and livelihood security for future agriculture.

Keywords: *Food, nutritional, livelihood security, agricultural interventions.*

1.1 Household Food, Nutritional and Livelihood Security-Definition and Concept

Food an individual eats fundamentally affects his health, strength, stamina, nervous condition, moral and mental functioning. It is of paramount importance in the normal growth, development and health of humans. The access to food by all is still unachieved but cherished goal. A widely accepted food security comprises of three parts.

- ☆ Every individual has a physical, economic, social and environmental access to balanced diet that includes necessary macro and micro-nutrients, safe drinking water, sanitation, environmental hygiene, primary health care and education so as to lead a healthy life.
- ☆ Food is produced from efficient and environmental friendly technologies that conserve and enhance the natural resource base of crops, animal husbandry, forestry, inland and marine fisheries etc.
- ☆ The ultimate composition of a food is expressed in terms of nineteen chemical elements.

Human being require carbohydrates, lipids, fatty acids, vitamins, proteins and several macro and micro elements for normal growth and development as well as sustaining good health. A list of essential nutrients required for sustaining human life is given in Table 1.1 Agriculture is a very important sector for the sustained growth of the Indian economy. About 70 per cent of the rural households and 8 per cent of urban households are still principally dependent on agriculture for employment. Green revolution succeeded in India to increase the farmer's income, yield of major crops and made India self-reliant in food production, with the introduction of high yielding varieties and use of synthetic fertilizers and pesticides [1]. In the post-green revolution period agricultural production has become stagnant and horizontal expansion of cultivable lands become limited due to burgeoning population and

Table 1.1: Essential Nutrients for Sustaining Human Life*[3]

Carbohydrate	*Lipids and Fatty Acids*	*Vitamins (Amino acids)*	*Micro-elements*	*Macro-elements*	*Protein*
Water Carbohydrate	Linoleic acid	A, D, E, K	Fe, Zn,	Na	Methionine
	Linolenic acid	C (Ascorbic acid))	Cu, Mn	K	Phenylalanine
		B1 (Thiamin)	I, F, B	Ca	Threonine
		B2 (Riboflavin)	Mo, Se	Mg	Cystine
		B3 (Pantothenic acid)	Ni, Cr	S	Cystiene
		Niacin	V, SI	P	Tryptophan
		B6 (Pyridoxal)	AS, LI	Cl	Valine
		Folate	S, N		Histidine
		Bioin			Isoleucine
		B12 (Cobalamin)			Leucine
					Lysine
(2)#	(2)	(13)	(17)	(7)	(11)

* Essential and beneficial substances in foods are also known to contribute to good health

Total number of essential items is required in each item.

industrialization. In 1952, India has 0.33 ha of available land per capita, which is now 0.15 ha [2]. As a result of increase in production through vertical dimension (yield per hectare basis) as horizontal become constrained, our agriculture become chemicalized through heavy use of synthetic fertilizers and pesticides which threatened the sustainability of our agriculture. In this situation, it is essential to develop eco-friendly technologies and concepts for maintaining crop productivity. The change in trend of population is given in Figure 1.1. The estimated impact of soil degradation in Indian agriculture is given in Table 1.2.

Table 1.2: Estimated Impact of Soil Degradation on Indian Agriculture

Crop	*Per cent Loss*	*Loss (M US$)*
Paddy	2.7-4.7	189
Wheat	3.6-6.4	248
Barley	4.5-7.0	8
Groundnut	2.8-4.4	110
Gram	5.6-7.8	60
Rapeseed and Mustard	5.8-8.5	155
Jowar	5.7-7.6	40
Bajra	6.8-8.4	25
Cotton	5.3-8.8	140
Maize	3.2-4.9	25
Sugarcane	4.5-7.9	200
Other crop	4.0-6.3	750
Total	4.0-6.3	1950

Source: Down to Earth 6 October, 1996.

1.2 Knowledge Management (KM) in Agriculture

Now there is a realization that knowledge and capital are the most determinant of food security and development of nation. Knowledge coupled with efficient resource management is now considered as a driven force for all over, social, economical and ecological development. Optimization of technology, human resources and energy inputs are essential for agricultural production

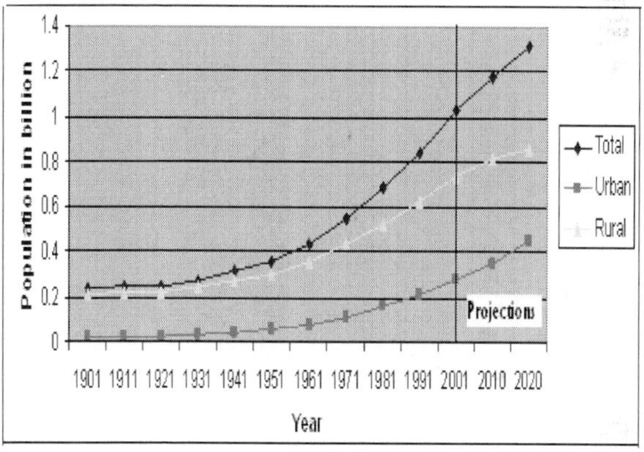

Figure 1.1: Changing Trend of Rural and Urban Population

with limited land and water resource. Instead of traditional factors of production-land, labour and capital knowledge, technology and social capital are today more important and crucial for enhancement of production and productivity in agriculture. The recent trends in food production in India and the average annual growth rate in agricultural sectors in India given in Tables 1.3 and 1.4.

Table 1.3: Recent Trends in Food Production (MT/yr) in India

Year	*Grain Production*	*Root Production (Grain equivalent)*	*Commercial Inputs*	*Food Aid (Grain equivalent)*	*Total Availability of all Foods*
1994	170.8	6.2	0	0.4	244.7
1995	174.9	6.1	0	0.4	249.7
1996	177.8	6.4	0.4	0.4	255.9
1997	182.8	7.8	1.3	0.3	261.3
1998	184.0	6.4	1.6	0.3	260.5
1999	191.0	7.9	1.4	0.3	267.4
2000	192.9	8.2	0	0.3	263.0
2001	197.4	7.6	0	0.2	278.8
2002	173.2	8.1	0.04	0.3	281.3
2003	189.5	8.3	0.3	–	279.5
Projection					
2008	216.3	9.0	0.3	–	316.5
2013	238.3	9.9	0.4	–	344.5

Table 1.4: The Average Annual Growth Rate in Agricultural Sector in India

Year	*Annual Growth in Agriculture and Allied Sectors (per cent/yr)*
1985–1990	3.2
1990–1992	1.3
1992–1997	4.7
1997–2002	2.1
2002–2003	6.9
2003–2004	10.4
2004–2005	0.7

1.3 Food and Agriculture Organization (FAO) Integrated Plant Nutrient Supply System (IPNS) Approach

Integrated plant nutrient supply system (IPNS) maintains soil fertility and plant nutrient supply by optimizing the benefits from all possible sources of plant nutrients. The key objectives of IPNS are:

- ☆ To maintain or enhance soil productivity through a balanced use of mineral fertilizers combined with organic and biological sources of plant nutrients.
- ☆ To improve the stock of plant nutrients in the soil.
- ☆ To improve the efficiency of plant nutrients, thus limiting losses to the environment.

Balanced plant nutrition help in healthy food production system. The apparent decline in returns from the increased fertilizers application include.

- ☆ Imbalance N, P and K application and the latter two are being applied in too low amounts in some states.
- ☆ Deficiencies of secondary and micro-nutrients are appearing with increasing frequency.
- ☆ More fertilizers are being used on soil inherently poorer in fertility and/or uncertain water supply as in dry land areas.
- ☆ The increased intensity of cropping together with changes in crop sequences of cereal rotations in place of cereal legume rotations.
- ☆ There may be an overall decrease in soil organic matter. Status of soils.
- ☆ Built up of certain disease, pests and weeds under intensive system of cropping.

To overcome the negative effects of application of plant nutrients, both at low or imbalanced and high levels of input can be avoided by excellent management. Balanced fertilization supplemented with organic nutrient sources help in overcoming the hazards of nutrient depletion and of mining soil fertility. IPNS provides excellent opportunities to overcome all the imbalances besides sustaining soil health and enhancing crop production. The concept of INM is the maintenance or adjustment of soil fertility and of plant nutrient supply to an optimum level for sustaining the desired crop production through optimization of the benefits from all possible sources of plant nutrients in an integrated manner.

1.4 Opportunities for Organic Farming (OF)

Organic farming largely excludes the use of chemicals. Presence of residue of toxic chemical compounds in food items has become a major concern for health all over the world. These compounds came into the food chain because of indiscriminate and over use of agro-chemicals (fertilizers, pesticides, hormones etc. during crop production, storage, processing and value addition. Substantial quantities of these toxic compounds like HCN, DDT, Edosulphan, chlorpyriphos etc. and heavy metals like lead (Pb), nickel (Ni), chromium (Cr), cadmium (Cd) and mercury (Hg). Stay in food items in residual form and come into the food chain. Consequently such compounds reach into human and animal bodies. Accumulation of such compounds in the body creates health hazards. Studies have established the presence of these toxic compounds in all types of food items including in animal products. Therefore, demand of food items which have no residue is increasing. Organically produced food items are toxin free and are tasteful. Presently the comparison in organic Vs conventional looks something like given in Table 1.5.

1.4.1 Organic Inputs

Organic inputs have long been regarded as being essential for the maintenance of soil fertility. It is native to this land. Apart from contributing modest quantity of primary nutrients, they also offer a good source of secondary and micronutrient essential for plant growth. Organic agriculture requires scientific management of organic inputs *e.g.* vermicompost, bio-inoculants, rock phosphate, farm yard manure (FYM) etc. with a systematic approach for their management. Organic manures, vermi-compost and green manure etc. have been used as a means of maintaining and improving soil health. A variety of organic inputs used in Indian agriculture system offer several advantages viz: improve soil physical and biological properties, protect water quality and save energy/low cost inputs, reduce health risk by protecting farmers from hazards of chemicals, support a true economy, promote biodiversity and improve quality of produce. Organic foods are better in appearance, flavour, taste and nutrition. There has been observed a gradual decline in some important mineral content (Ca^{+2}, Mg^{+2}, Na^{+}, K^{+}, P, Fe^{+2}, Cu^{+2}) in crops over the last sixty years. This can be attributed to higher synthetic inputs Intergrated use of different organic inputs is required depending on their availability and crop pattern for higher and sustainable crop production. Comparison of organic and inorganic farming system given in Table 1.5.

Table 1.5: Comparison of Organic and Inorganic Farming System[4]

Parameter	*Organic*	*Conventional*
Size	Smaller, marginal, dependent operations.	Large scale, economically tied to major food corporation
Method	No use of purchased fertilizer and other inputs *e.g.* pesticides, weedicids etc. less mechanization of the growing and harvesting process. Use of organic inputs like green manure, vermicompost, bio-fertilizers etc.	Heavy use of chemicals *e.g.* fertilization, use of pesticides etc. mechanized production using special equipment and facilities.
Technology	Nature based, environment friendly and sustainable.	Synthetic, harmful to environment and nutrient depleting.
Products	Good in taste, flavour, nutrition and free from chemicals.	Tasteless, less nutritions, may contain toxic residues of chemicals.
Market	Local direct to consumer, on farm stand and farmers market and through special wholesalers and retailer.	Wholesale with products distributions across large areas average supermarket produce travel 100 to 1000 km and sold through high volume.

Organic farming is visible in peri urban areas in vegetables crops. The comparison of yield of vegetable in India and China is given in Table 1.6.

1.5 Role of Biotechnology in Sustainable Crop Production and Food Security

Plant biotechnology has emerged as a supplemental tool to increase the efficiency of crop production by way of developing transgenic plants with improved traits *viz.* as disease resistance, insect/pest resistance, herbicide tolerance, abiotic stress tolerance, nutritional quality improvement, delayed fruit-ripening and improvement in keeping quality. The transgenic technology has emerged as a supplemental tool to increase the efficiency of crop plants with improved traits. Though traits governed by single or a couple of genes can be easily manipulated through the transgenic routes and solution of the majority of the other problems of biotic and abiotic stress can be obtained. Hence plant bio-technology will be future of sustainability of food grains.

Table 1.6: Comparative Yields (t/ha) of Vegetables in India

Year	*India*	*China*	*World*
1990	10.2	17.7	14.9
1995	10.2	18.8	15.5
2000	13.1	18.9	16.6
2002	12.5	19.6	16.9
2003	12.9	19.2	16.8

1.6 Efficient Use of Renewable and Non-renewable Inputs in Agriculture

India's irrigated agriculture has 60 million ha of net irrigated area. The input use efficiency in this kind of agriculture continues to be quite low. This is particularly true of use of irrigation water and of chemical fertilizers. Thus, the irrigation water use efficiency in rice may be as low as 30 per cent. This is also true of applied fertilizer nitrogen. Thus, while India has developed the largest irrigation system in the world and a nitrogen fertilizer industry which is believed to be the fourth largest, the use of these costly inputs tends to be highly inefficient. This results in increased cost of production also in sustainability and environment problems. Further, India makes very little use of renewable resources of inputs in its agriculture. The efficient use of renewable and non-renewable inputs in crop production contributes a significantly in food security. The sector wise energy conservation statistics for India is given in Table 1.7.

Table 1.7: Sector-wise Energy Conservation Statistics for India[5]

Sector-wise Demand in (per cent)		*Sector-wise Energy Saving Potential (in per cent)*	
Agriculture	5	Agriculture	up to 30
Industry	50	Industry	up to 25
Transport	22	Transport	up to 20
Domestic	10	Domestic	up to 20
Others	13	Economy as a whole	up to 23

1.7 Nutrient Management (NM)

Day to day decline in inherent capacity of soil to produce crops is now a serious concern. Out of N, P and K, deficiency of nitrogen is universal. It is difficult to retain it in the soil due to various loss mechanisms. Most of Indian soils are low to medium in phosphorus. With the passage of time, the potassium deficiency has also become widespread. Special attention should be paid to sulphur nutrition, as agriculture production intensities, high analysis fertilizers having little or no sulphur are increasingly used. Unbalanced fertilization has also resulted in micro-nutrient deficiencies. The approaches to arrest micro-nutrient depletion include matching the crop removals of the micro-nutrients with additions through respective external carriers, supplementation through organic sources mobilization and transformation through growing micro-nutrient efficient crops and crop cultivars. Farmers are in lack of fertilizer management practices hence loss of nutrients during application has been increased. Farmers using a variety of fertilizers, pesticides and chemicals. The application of fertilizers on the basis of soil test based recommendation is of great advantage in boosting food grain production.

1.8 Good Quality Seeds/Planting Materials (Genetic Materials)

Seed is the most critical input on which the efficiency of all other inputs depends. Every efforts should be made to improve the seed replacement rate from the present 12 per cent to at least 20 per cent. The seed production by the government can be limited to the production of breeder and foundation seed and inbred lines for hybrid seed production. Private sector should be encouraged in seed multiplication programme along with the involvement of farmers either Govt. or Non-Govt Organizations (NGOs) for better harvest.

1.9 Use of Precision Farming (PF)

PF has been getting attention by many concerned with the production of food, feed and fiber and sustainability and efficiency of agricultural system. While discussing the new gains to be made towards sustainability and food security advised, that at the production level, precision farming practices will have to be developed and popularized [6]. Precision farming (PF) technology will be used for three quite different purposes. First, this technology will enable farmers to do better more accurately what they are already doing. For example, GPS can provide swath guidance to reduce overlap and skips or to all night operations. Second, PF information will allow farmers to evaluate their current practices and experiments. This is likely to be the most immediately useful aspect of PF for many farmers. Third may farmers will elect to vary some practices across the field. The most obvious candidate for this is chemical application, but a thoughtful farmer may elect to very other actions such as tillage or seeding. The tremendous quality of information to be handled in a PF application would not be manageable without a computer. This proper software for the job is a "Geographic Information System" (GIS). This software keeps track of PF data by referencing each measurement to its location. Think of it as computer map management system.

1.10 Information Technology and Computer (IT &C) in Agriculture

The information systems are computer based information banks capable of providing information on some specific field/areas of application. In agriculture, it can be developed on various crops, diseases, insects and cultural practices etc. The potential of information systems (IS) can be assessed broadly under two heads: (a) as a tool for direct contribution to agricultural productivity and (b) as an indirect tool for empowering farmers to get informed and take quality decisions which will have positive impact on the way agriculture and allied activities are conducted.

The benefits of IS in empowering Indian farmer are significant and remains to be exploited. The Indian farmer urgently requires timely and reliable sources of information inputs for taking decisions. Technologically, it is possible to develop suitable information systems to cater the information needs of Indian farmer. Use-friendly information systems, particularly with content in local languages, can generate interest in the farmers and other working at the growth.

1.11 Soil Resource Management (SRM)

The soil in India is suffering from varying degree of degradation. The marginal or problematic soil occupies about 57 per cent of geographical area of which water erosion alone constitutes 148.9 mha area (45 per cent). This results in an annual loss of 5333 mt of fertile soil which is equivalent to about 16.4 t/ha/year. The erosion depletes soil depth, fertility, organic matter, ground water table and consequently the productivity. Wind erosion causes loss of top soil is 1.9 per cent, terrain formation in 1.2 per cent and over lowering and shifting of sand dune is 0.5 per cent area of the country. The practice of shifting cultivation in north-east is a serious concern for land degradation which has to be taken seriously. Hence soil management particularly soil and water management is to be given priority for food sustainability.

1.12 Bio-fortification of Crops

Micronutrient deficiencies affect three billion people worldwide. Malnutrition hinders the development of human potential and nation's social and economic development. The World Health Organization (WHO) and Consultative Group on International Agricultural Research (CGIAR) have made fighting the hidden hunger (*i.e.* micronutrient deficiencies) a high priority. The micronutrients, iron, zinc and vitamin A have been targeted for intervention due to immense magnitude of the problem amongst the world's poor. Emphasis of WHO is on supplementation and fortification. CGIAR efforts focus on biofortification through its Harvest Plus Programme. Micronutrient content of the staples of the poor (rice, wheat, maize, beans, cassava and sweet potatoes) is being improved through breeding and biotechnological approaches. Excellent example of application of biotechnology application is the development of so called "golden rice" with adequate levels of β-carotene as all the existing rice varieties lack β-carotene. Africa Harvest program is developing "super sorghum" with the aim of improving digestibility, and increasing the level of Vitamin A and E and iron and zinc [7]. Bio fortification provides excellent opportunities to tackle the problems of malnutrition as well as food security.

1.13 Policy issues

1. Strengthen land conservation authorities and community groups at the national and local level and encourage cooperation between stakeholders.
2. Develop national action plan to protect biodiversity, combat land degradation, focusing on the most important threats to biodiversity and causes of land degradation.
3. Institute legal, institutional and economic policy reforms ensuring cross-sectoral coordination to combat desertification. Integration and coordination of national action plan, MOEF and ministry of agriculture regarding achieving conservation efforts.
4. Provide appropriate incentives to small farmers, pastoralists, and communities for better land-management practices.
5. Promote rural credit and mobilize rural savings through the establishment of rural banking systems.
6. Develop infrastructure, as well as local production and marketing capacity.
7. Strengthen the capacity of national institutions to analyze environmental data and monitor environmental change, including systems to provide early warning of droughts.
8. Regulate use of groundwater and policies for low water requiring crops, micro-irrigation and recommended land use policies.
9. Implement accelerated afforestation and reforestation programmes using drought resistant, fast-growing species.
10. Intensify production on appropriate lands to reduce the need to cultivate marginal lands. Fallow–farming system approach by crop diversification etc.
11. 'Crop and animal insurance' against crop failure due to drought or causalities of animal due to the epidemic.

12. Use of modern tools is imperative in future agriculture.
13. Integrated plant nutrient supply system (IPNS) is vital for crop productivity and soil health management must be adopted at farm levels [8].
14. Subsidy policy must be reformed. The benefit should directly go to farmers rather than to companies.

1.14 Conclusions

The new era of agriculture is challenging in term of resource degradation. Hence one must convert challenges into opportunities for food and nutritional security by adopting new interventions coupled with traditional wisdom of old traditional agriculture. The factors of production like, integrated nutrient management, use of precision farming, biotechnology, knowledge management, soil resource management, quality, watershed management etc. are of critical importance. All stakeholders either scientist, farmers, companies (public or private) should work strategically to combat the challenges of second generation agriculture for ensuring food and nutritional security.

1.15 References

1. Ghosh, S.K., Murthy, K.M.D., Ramesh, G. and Palanaiappan, S.P. (1999) Green revolution and its impact on Indian Agriculture Employment News 222: 1-2.
2. Singh, K.K. and Shekhawat, M.S. (2000) Precision agriculture and information technology. *Farmer and Parliament*, 10-13.
3. Welch, R.M. (2004) Micronutrient, agriculture, and nutrition for improved health. In IFA International symposium on micronutreint. Feb. 23-25, 2004, New Delhi.
4. Subhash Chand, Tahir Ali and K.A Sofi (2006) Food Security: Today's Concern and tomorrow's Prosperity. Journal of Agricultural Research & Development, 2: 1-6.
5. Kembhavi, P. (2008) Energy conservation and energy management, Green Energy: 4: 2, pp. 36.
6. Swaminathan, M.S. (2002) Food security and sustainable development. In Acharya, S.S., Surjit Singh and Vidya Sagar

(ed), Sustainable Agriculture, Poverty and Food security: agenda for Asian economics Vol. I pp.11-32.

7. Khush, G.S. (2008) Biofortification of Crops for Reducing Malnutrition, Proc. of Indian Science Academy. Delhi.
8. Subhash Chand (2008) Integrated nutrient management for sustainable agriculture, IBDC, Lucknow, pp112.

Chapter 2

Rural Agriculture Development for Food and Nutritional Security

Bread is God to hungry people

—*Mahatma Gandhi*

Every human being needs healthy and nutritious food for livelihood. Fostering the growth of national and global food supplies is necessary for eliminating hunger and reducing poverty, but it is not enough. Today, even in the midst of sufficient global food supplies, 800 million people are hungry because they cannot afford to buy the food they need for a healthy life. More than two billion people are at risk from micronutrient deficiencies (*e.g.* vitamin A, iodine, iron, zinc, manganese etc.), and more than 1 billion are actually disabled by them–harmed by mental retardation, learning problems, and blindness. Ironically, nearly 75 per cent of poor and undernourished people live in rural areas where food is grown. Reducing poverty and hunger will require encouragement of rural development in general and a prosperous small-holder private agricultural economy in particular. Encouraging rural development is the best way to help poor farmers and rural dwellers become more productive and improve their living standards. It is also critical to increasing national and global food supplies. Further, rural development can contribute significantly to improved food and nutritional security of the country as well as world.

2.1 Assuring Food Security

The job of assuring food security is large and complex. Action needs to be taken simultaneously at the household, national, and global levels to achieve the following goals.

2.1.1 Increase Agricultural Output Worldwide

Over the next 30 years, developing countries' food needs could nearly double because of population growth and modest income growth. The changes in man and arable land ratio over the years given in Figure 2.1. and population living below the poverty line given in Figure 2.2.

2.1.2 Reduce Poverty

The best way to reduce poverty and hunger is through economic growth and indeed, few countries have significantly reduced poverty without it. For most developing countries, improved agricultural productivity can be the engine of non-agricultural growth.

2.1.3 Improve Health and Nutrition

Eliminating hunger requires targeted nutrition, health and food programs. Increasing family income alone does not ensure that people will consume the right kind of nutrients in the right quantities at the right times to maintain their health and productivity. Today, most households could prevent child malnutrition if they used existing resources optimally, making small changes in their health and nutrition behaviour. Improving diets often requires nutrition counseling, prenatal nutrition services, and public health interventions. In some places, it also requires investments to correct micronutrient deficiencies. These often cost little but generate large returns. Thus, although general poverty, infrastructure, and agriculture programs will improve nutrition eventually, direct actions are likely to have faster and greater impacts. The challenge of assuring food security is significant and needs attention now. It cannot be met without renewed commitment by scientists, farmers, national policymakers, international donors, and the World Bank to increase agricultural productivity through research and technology development and to implement policies and programs

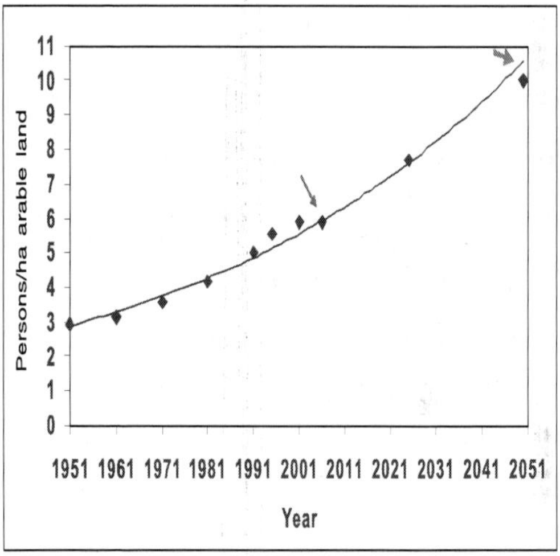

Figure 2.1: Changes in Man-Arable Land Ratio Over the Year

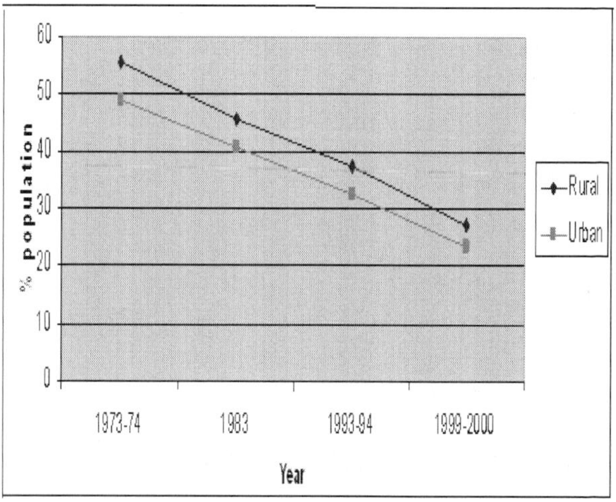

Figure 2.2: Population Living Below the Poverty Line

that will ensure that the poor and hungry benefit from increasing agricultural productivity.

2.2 Past Progress and Outlook

During the past twenty-five years, substantial progress has been made in improving the living standards of people in the developing world. The proportion of the world's people living in poverty has declined; per capita incomes have doubled; infant mortality has fallen by half; and average life expectancy has increased by ten years since the 1970s. In addition, global agricultural productivity has risen sharply; total calorie supplies per person have risen by 30 percent; and real food prices have fallen by more than 50 percent. The increase in productivity has allowed consumers to improve their diets in terms of calories and mineral nutrition. A photograph of farmer weeding in aerobic paddy is given in Plate 2.1 showed that farmers are eagerly waiting for new emerging technologies for high yield sustainable agriculture.

2.3 Encouraging Appropriate Policies and Strategies

Developing countries like India and China need to implement sound and stable macroeconomic and sector policies. They are increasingly recognizing that heavy government interference in the productive activities of their agricultural economies has inhibited

Plate 2.1: Farmer Weeding in Aerobic Paddy

agricultural growth and distorted the allocation of resources. Through analytical work, policy dialogue, and financial support, the Bank is assisting countries in liberalizing prices of farm commodities and inputs, reforming public enterprises, liberalizing agricultural trade, and changing foreign exchange and taxation regimes which discriminate against agriculture. Even if world food supplies grow dramatically over the next 30 years, fast-growing countries such as China could become major importers. If they and other countries are to refrain from costly food self-sufficiency policies, they must be guaranteed stable, long-term access to world markets. Reducing the import restrictions of the rich industrial countries is also critical to increasing the demand for the agricultural products of developing countries, which can help considerably in generating employment and reducing poverty there. The World Bank is actively promoting greater access to rich country markets for the agricultural and agro-industrial products of its client countries and is supporting actions in the World Trade Organization to achieve this objective. The farmer weeding in aerobic paddy is given in Plate 2.1.

2.4 Enhancing Food Supplies through Intensification of Production Systems and Through Sound Natural Resources Management

2.4.1 Encouraging Rapid Technological Change

Implementing rapid technological change on the hundreds of millions of farms in the developing world is essential for agricultural and income growth. Investing in the research necessary to stimulate technological change in agriculture is a high priority for the Bank. Each year, it lends more than $220 million to national agricultural research institutes, in addition to contributing more than $45 million per year to the Consultative Group on International Agricultural Research (CGIAR). The Bank is supporting research on crops and processes that are of little interest to the private sector, but which could have a large impact on rural poverty and hunger; these include subsistence crops and crops that are staples in poor regions, such as maize, cassava, sweet potato, millet, and sorghum. It is also working with the international community to ensure that the poorest communities in developing countries will be able to benefit from the breakthroughs in technology that are increasingly being generated and patented by the private sector.

2.4.2 Increasing the Efficiency of Irrigation

Irrigation accounts for 70 per cent of the fresh water used by man and has contributed greatly to the production increases seen during the twentieth century. However, agriculture is increasingly competing for water with urban and industrial users. There will be sufficient water for all only if agriculture and other sectors greatly improve the efficiency of their water use. This will require improving incentives to water users by establishing water markets, clarifying water rights, and pricing water to reflect its true value. The Bank is assisting countries to improve the efficiency of irrigation systems as part of their comprehensive water resources planning. The plan wise irrigation potential created and utilized in India (Cumulative) is given in Figure 2.3.

2.4.3 Improving Natural Resource Management (NRM)

The World Bank is involved in many projects that support the intensification of agriculture and, at the same time, encourage better natural resources management. A community-based approach to resource allocation, enforcement, and maintenance has proven successful in such diverse locations as Burkina Faso and northeast Brazil; it is now being adopted in Egypt and is being incorporated into many new agricultural development projects elsewhere. For example, social forestry projects are under way in Asia and Africa. And a major watershed rehabilitation project is under way in the Loess plains in China: slope lands are being terraced; orchards and grasslands are being planted; and sediment control dams are being built. This work has enabled farmers to double their crop output while significantly reducing soil erosion. The Changing trends in productivity of crops grown in irrigated and rain fed areas is given in Figure 2.4.

2.4.4 Improving Access to Food

2.4.4.1 Strengthening Markets and Agri-businesses

The support of markets and agribusiness has received insufficient attention in the World Bank's assistance to agriculture and rural development, with the exception of assistance provided by the International Finance Corporation. This is now changing, as the power of markets to efficiently allocate resources–and reduce price margins between consumers and farmers–becomes

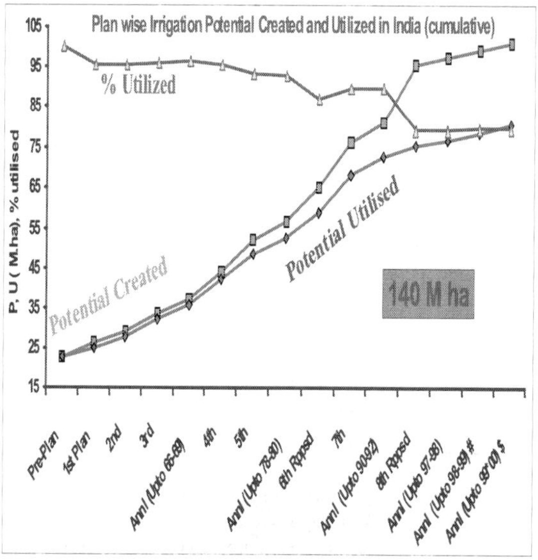

#-Anticipated, $-Target

Source: CWC (2002)

Figure 2.3: Plan-wise Irrigation Potential Created and Utilized in India (Cumulative)

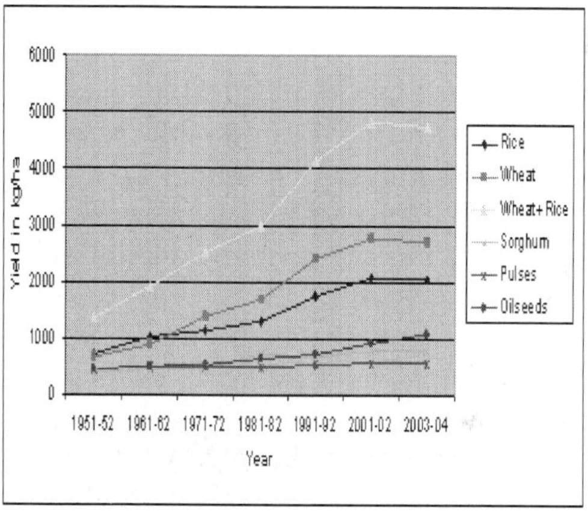

Figure 2.4: Changing Trends in Productivity of Crops Grown in Irrigated and Rainfed Areas

increasingly accepted. Where the state either has withdrawn or is withdrawing from marketing and input supply–as in Eastern Europe and Central Asia, Africa, and Latin America–the Bank is assisting governments both to develop the legal, financial, and institutional frameworks that are necessary for competitive markets to work and to establish information systems for collecting and disseminating vital data.

2.4.4.2 Providing Education and Health Services to Both Boys and Girls

Providing education and health services to both girls and boys is one of the key ways to reduce poverty and hunger. There is substantial evidence that individuals' education is closely linked to their incomes and that improved education contributes to national economic growth. Education and health services are especially important for women, who have a major role to play in growing crops and in reducing hunger. Better-educated and healthier women are much more productive and earn higher incomes. Since women often use their additional income on investments in family welfare, increases in their incomes are likely to have greater immediate and long-term impacts on poverty and hunger than equal increases in men's incomes. Education for girls also lowers fertility rates and improves environmental management. For increasing population of India, it is mandatory to provide better health services and education to boys and girls for healthy young generation. The changing decadal population trends in India are given in Figure 2.5.

2.4.4.3 Investing in Infrastructure

When there are adequate communications networks, roads, storage facilities, supplies of electricity farmers can obtain the information they need to grow the most profitable crops, store them, move them to market, and receive the best price for them. Today, up to 15 per cent of production is lost between farm gates and consumers owing to poor roads and storage facilities, reducing farmers' incomes and raising urban consumers' food costs. As cities grow, the need for infrastructure becomes all the more important. Helping countries build the infrastructures they need has long been a core World Bank activity, and it continues to be so today. Along with providing investment capital for infrastructure, the Bank is helping countries develop rural infrastructure strategies that include clearly articulated

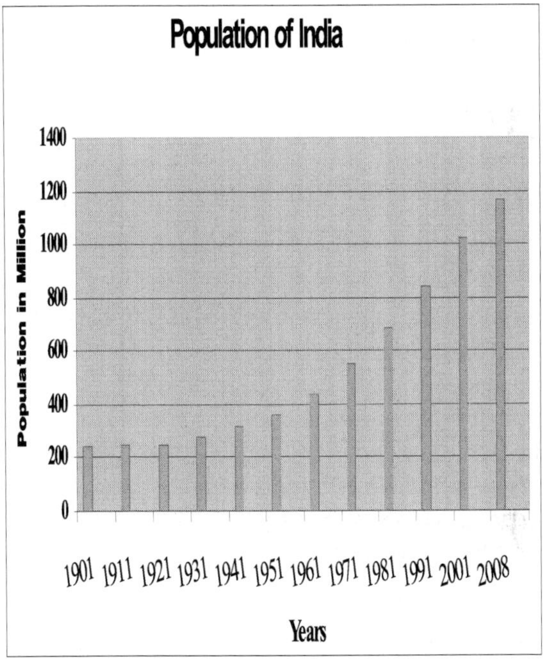

Figure 2.5: Changing Decadal Population Trends in India

priorities and are founded on strong analysis of costs and benefits. It is also helping countries design new approaches to financing that utilize private sector and local resources.

2.4.4.4 Fostering Broad Participation

Experience shows that development projects are much more likely to reflect the affected community's priorities, reach their goals, and be sustainable when they are designed and executed with a high degree of influence by local stakeholders. The Bank is assisting communities and local governments to find ways to finance infrastructure and services using their own revenues and fiscal-transfer mechanisms, develop their legal authority, strengthen their administrative and technical competence, and develop participatory mechanisms for assessing projects.

2.4.5 Improving Food Utilization

2.4.5.1 Integrating Household Food Security and Nutrition Policy into Rural Development Operations

Working with its partners, the World Bank is deeply engaged in supporting efforts to reduce hunger and malnutrition. It has made particularly strong progress in helping developing countries improve their poor populations' nutrition, for which analytical work and financial support have expanded rapidly over the past several years. And it is now systematically integrating food-security concerns into agricultural policy dialogue and reform, and incorporating nutrition projects into Bank operations in other sectors, such as agriculture, education, adjustment, and safety net operations and sector investment loans.

2.5 Conclusions

For ever increasing population, India needs to produce 300 Mt food grains in 2020 from limited arable land holdings (145 mha.) to feed the nation. The food security besides good nutrition is of prime concern. For rural sustainable agricultural development,it is necessary to utilize all the natural resources effectively without land degradation coupled with modern agro techniques (resource conserving technologies, precision farming, INM, organic farming, watershed management, rain fed farming etc). The rural agriculture development of an area decide the overall well being of human kind.

2.6 References

1. Subhash Chand (2008) Integrated nutrient management for sustaining crop productivity and soil health. IBDC, Lucknow. pp112.
2. Tim Dyson, (1996), Population and Food: Global Trends and Future Prospects (New York: Routledge).
3. Food and Agriculture Organization (1993) The State of Food and Agriculture (Rome).
4. Food and Agriculture Organization/World Health Organization, 1992, International Conference on Nutrition and Development–A Global Assessment (Rome).
5. World Bank (1994) Enriching Lives: Overcoming Vitamin and Mineral Malnutrition in Developing Countries, Development in Practice Series (Washington).

2.7 Websites

1. www.foodsec.org/docs/concepts_guide.pdf
2. www.canadainternational.gc.ca/g8/ministerials-ministerielles/2009-04-2004-AgMin-Pres.aspx
3. http://csdngo.igc.org/agriculture/agr_How_to.htm
4. www.adb.org/Media/InFocus/2009/agriculture.asp)

Chapter 3

Agricultural Production Based Food Securities

Ensuring food security and reducing rural poverty are indispensable elements of attaining inclusive and sustainable economic growth in Asia.

—Katsuji Matsunami
Practice Leader for Agriculture, Food Security, and Rural Development

The healthy and nutritious food is a primary right of every mankind besides livelihood security. Agriculture sector and allied activities contributing about 22 per cent in GDP in our national economy and sustaining life of >65 per cent mankind. Production based livelihood security is an important area having capability of providing ample employment opportunities for living a healthy life. The country like India and China have good manpower for production based systems. Agricultural production systems provide livelihood security without high risk of health at village level without any specific technical qualifications. However, modern intensive agriculture is skilled oriented, enterprises activity needed less initial cost like any other industrial production systems. Besides marginal farmers with small land holding (<0.2 ha) in Indian agriculture, is continuously providing livelihoods to a major population in villages and towns. Food security at the macro level results from augmenting food production; food security at the household level can come from both increasing household food production as well as by increasing the household income through expanding and strengthening its other livelihoods options and improving market access to the produce from households. Augmentation of food production in turn occurs through improving productivity of land through watershed development, irrigation, improvement in agronomic practices, introducing new crop combinations, cropping techniques and soil health.

Keywords: *Production, livelihoods, food security, Indian agriculture*

Efficient and equitable use of water is important for food and ecological security. Working towards ecological security includes biodiversity conservation, strengthening technical and social measures for regeneration of fragile ecologies and contributing towards a more prudent and sustainable use of natural resources. India's livestock sector has been growing at an impressive rate of about 5 to 6 per cent a year. Livestock sector plays a significant role in supplementing family income and generating gainful employment in the rural sector, particularly among the landless, marginal and small farmers and women, besides providing cheap nutritional food to millions of people. Livestock has been emerged as an integral part of the farming systems in India, since time immemorial. It is pursued as a subsistence occupation to agriculture mainly to provide food for human consumption, draught power and obtain dung manure for crop production. Livestock rearing is closely integrated with crop raising activities. There is a need to enhance the productivity of livestock sector and crops through appropriate technological interventions to ensure transformation of livestock sector towards commercialization. Poor people who live near forests, for whom forests and trees have, traditionally, been an important source of livelihood. Over the years, the implementations of protection-oriented public policies have made serious inroads to these livelihood activities, resulting in decreased incomes for the local communities. A Carbon Neutral Bio-diesel Fuelled Energy System (CNBFES) need to be developed as a village scale technology using local forest based oil-bearing species such as Mohua, Karanj, Neem, Jatropha, Apricot, and agricultural oil seeds such as Niger, caster etc. Production of fuel woods, timbers, fodders cocoons for silk, honey etc. are the source of poor people livelihood. Scientists and extension workers should be coordinate through suitable technologies for sustainable livelihood to fulfill the needs of the poor farmers and local forest dependent people.

3.1 Various Types of Production Systems

There are several agricultural based production systems described hear under following headings.

3.1.1 Crop Production Systems

Major crops are cereals, pulses, oilseeds, fiber, forage and medicinal and aromatic plants etc. These systems contribute in met out food, fiber, fuel, fodder needs of majority of growing human and livestock population of the country. The agricultural production scenario in India is given in Table 3.1a and 3.1b.

Table 3.1: Agricultural Production Scenario in India [1]

Crop	Production (Million Tonnes)		
	2003–04	2007–08	% Increase/Decrease
Rice	88.5	94.1	6.3
Kharif	78.6	81.5	3.7
Rabi	9.9	12.6	27.3
Wheat	72.2	74.8	3.6
Coarse Grain	37.6	36.1	-3.9
Kharif	32.2	28.6	-11.2
Rabi	5.4	7.5	38.9
Pulses	14.9	14.3	-4.03
Kharif	6.2	5.8	-6.5
Rabi	8.7	8.6	-1.15
Total Food Grains	213.2	219.3	2.9

3.1.2 Medicinal and Aromatic Plants Based Production System

In the diversification era of agriculture, growing of medicinal crops in the existing cropping systems would be more appropriate to boost up farmers income and fulfill the nation's domestic and export demand. The global demand of such products is rising

Table 3.1b: Production, Productivity Scenario in Jammu & Kashmir

S.No.	Crops	Area (000 ha)		Production		Production/ha (yield) (g)	
		2006–07	2008–09	2006–07	2008–09	2006–07	2008–09
1.	Rice	252.52	257.63	5546	5637.92	21.96	21.88
2.	Maize	323.60	315.80	4869	6331.72	15.05	20.04
3.	Wheat	266.11	278.72	4983	4835.63	18.73	17.43
4.	Pulses	29.06	29.99	141	138.83	4.85	4.63
5.	Other Cereals and Millets	44.26	40.29	238	227.77	5.38	5.65
	Total Food Grains	**915.55**	**922.43**	**15777**	**17171.87**	**17.23**	**18.62**
6.	Oilseed	64.30	65.24	413	495.90	6.42	7.60
7.	Fruits and Vegetables	83.95	89.64	NA	480.42	NA	5.36
8.	Condiments and Spices	2.71	2.69	24	24.57	8.86	9.13
			157.57		**1000.89**		**6.35**
	Total	**1066.5**	**1080.0**	**16214**	**18172.7**		**16.83**

substantially, therefore, appropriate agro-techniques need to be developed and standardized. Nutrient management particularly in cropping system is an ideal and most important crop production technology but proven research findings are very few for medicinal plants which can be adopted. The advantage of nutrient management in cropping system as compared to individual crop basis is more suitable and beneficial to rural livelihood. Therefore, present research work in medicinal plant based cropping system is proposed to select crops in cropping system which are sustained nutritional status of soil for longer time. India ranks second amongst the exporting countries, after China over 10,000 manufacturing units with a turnover of around Rs. 6200 crores/year. Projection for new aromatic industry in 2010 is given in Table 3.2. The field distillation unit and Patchauli are given in Plates 3.1 & 3.2. The Citronella, Lemon grass and *Bacopa monnieri* are given in Plates 3.3–3.6.

Table 3.2: New Aromatic Industry in 2010 (Projected) Perfumer[2]

Essential Oil	*2005 Production in Tons*	*Estimated 2010 Production in Tons*	*Value : million US$*
Essential oils (existing)			
Basil	500	1000	5.682
Cedar wood	150	200	0.909
Citronella java	400	550	3.375
Jamrosa	100	400	2.727
Lemongrass	400	800	8.182
Mentha citrata	50	150	1.875
Mentha piperita	800	1500	20.455
Mentha arvensis	18,000	28,000	254.545
Palmarosa	80	160	2.182
Pepper	80	120	10.909
Sandalwood	40	60	75
Spearmint	800	1300	14.773
Others	400	800	10.909
Total	21,800	35,040	411.523

Contd...

Table 3.2–Contd...

Essential Oil	2005 Production in Tons	Estimated 2010 Production in Tons	Value : million US$
Essential oils (new)			
Cymbopogon (new)	50	100	1.364
Geranium	50	100	9.091
Lavandin	10	50	1.364
Lavender	5	10	0.454
Patchouli	30	60	2.045
Ocimum (new)	50	100	1.364
Rosemary	20	40	1.364
Others	50	100	2.273
Total	265	560	19.319

3.1.3 Vegetable Based Production System

India is the second largest producer with 15 per cent of the total world production next to China. The average productivity of vegetables in India is 15 t/ha, which is low compare to developed countries. Per capita availability is just only 155g/day as against 285 g/day. In India as many as 65 vegetables are grown which have tremendous potential to increase production and fulfill the per capita need of vegetable. The various agro-climatic conditions are available in India, favour variability of fresh vegetable production through out the year for livelihood. The photographs of vegetables (radish, fenugreek & spinach, sugar beet and bean) are given in Plates 3.7–3.10.

3.1.4 Fruit Based Production System

The world fruit production is 370 M T. India ranks first with production of fruits is about 32 MT accounting 8 per cent and per capita availability is 100 g/day [3]. In total citrus accounts for 20 per cent of the world's fruit production. About 20-22 per cent of the total fruits are lost due to spoilage at various post harvest stages. This spoilage reduces the per capita availability to 80 g/day which is half the requirement for a balanced diet. Processing- Juice and concentrates, canned fruit, dehydrated fruit, Jams and Jellies etc. In India < 4 per cent of the fruits are processed as against 70 per cent in

Brazil, 60-70 per cent in USA, 83 per cent in Malaysia and 50 per cent in Israel.

There is need to develop fruit belt in different agro-climatic zones, identification of fruits/varieties in the light of changing environment/weather conditions of the fruit belts and keeping in view the market requirement. In apple traditional and spur verities have been

Plate 3.1: View of Field Distillation Unit

Plate 3.2: Patchauli

Plate 3.3: Citronella

Plate 3.4: Lemon Grass

identified. Spur varieties may be promoted in high mid hill 1500-2500 m. and delicious group may be promoted in high hills above 2500 m. Farmers with fragmented holdings may develop orchards on whole village/community basis. This will leads to higher production, better market opportunity in domestic and international market. The photograph sowing some tropical, subtropical and

Plate 3.5: ***Bacopa monnieri***

Plate 3.6: Rose Cultivation in Hi-tech Polyhouse

temperate fruits. The photographs of fruits are given in Plates 3.11–3.16.

Plate 3.7: Radish

Plate 3.8: Fenugreek and Spinach

Plate 3.9: Sugarbeet

Plate 3.10: Beans

Plate 3.11: Mango

Plate 3.12: Litchi

Plate 3.13: Banana

Plate 3.14: Apple cv. Red Delicious

Plate 3.15: Pineapple

Plate 3.16: Winery Yard (Grape orchard)

3.1.5 Ornamental Plants Based Production Systems

Floriculture, till recently considered to be a simple garden activity to get flowers for religious offerings and home decoration has emerged as an important agribusiness enterprises. India is among the important agriculture based economy in the developing world and has in the recent years recognized the importance of floriculture as a segment of agribusiness. With more than 86,000 hectare India is the leading country for floriculture in terms of area. Flower growing is practiced on very small holdings in most part of the country, perhaps large share of floriculture area goes unreported but, does this large area devoted to floriculture activities provides any significant edge to us burgeoning floriculture trades in the world estimated to be worth 50,000 crore US dollar. Floriculture in India has a long tradition; it has served the purpose of meeting our socio-cultural requirements since time immemorial. However, with rapid commercialization of agriculture and graduation of farming from subsistence level to commercial level, exposure to newer markets and opportunities have resulted in market segmentation and evaluation of niche market for sustainable rural production based livelihood in villages and socioeconomic development through flower marketing.

3.2 Forest Based Production System

Rangeland and pastures, woodland, social forestry, agro-forestry and recreational forestry systems comes under this system[4].

Multipurpose tree species planted in the arboretum are described below.

- ✰ *Acacia auriculiformis and Morus alba* suitable to be grown in degraded land can supply the fuel wood demand of the country.
- ✰ High value timber producing tree species as identified are *Gmelina arborea, Albizzia procera, Samania saman, Michelia champaca, Dalbergia sissoo, Eucalyptus hybrid,* and *Tectona grandis.*
- ✰ *Azadirachta indica* is an important source of several medicinal properties also provides good quality of furniture wood fodder leaves for rural livelihood.
- ✰ *Leucaena leucocephala* could be used to supply nutritive fodder for ruminants.

- ☆ *Gliricidia maculata*, as it is easy to propagate could be effectively used as live fencing and could supply foliage for fodder, mulch and manure.
- ☆ Silvi-horti or pastoral systems are highly remunerative and introduction of black pepper in a 3-tier system is definitely a value addition to utilize agro-forestry for spice production.
- ☆ Organic matter build up could be built in soils of agro-forestry to enhance soil productivity for longer time.
- ☆ Control of soil erosion and moisture retention could be two important facets of agro-forestry interventions for conservation of natural resources.

3.3 Livestock and Dairy System

The issue of sustainable agriculture and food security is frequently viewed in the narrowest of ways; of the technical and physical viability of the food production system. This system is, however, dependent upon factors which require broader considerations to be taken note of to ensure sustainability; people and their needs. It is thus not sufficient to design an agricultural system that ignores the needs for food of all of the population, including the producers themselves. Such an approach ignores the fact that the sustainability of the system is dependent upon not just equity across generations but within generations. The characteristics of small scale dairy management system is given in Table 3.3

The development of livestock production has been receiving significant priority as well as research attention in India in the last two to three decades. In the wake of various development programs the milk production has raised from 31.6 million tones to 60.8 million tones and egg production from 10 billion to 24.4 billion between 1980-81 and 1993-94 and the gross value of livestock output (in constant 1980-81 prices) from Rs. 7,387 crores to Rs. 17,937 crores between 1970-71 and 1990-91. Animal husbandry in India is an integral part of crop agriculture yet it's contribution to GDP is increasing in comparison to the contribution of agriculture, which is on decline. It contributes an estimated 8.4 per cent of the country's GDP and 35.85 per cent of agriculture output at current prices. It is main source of milk, meat, skin, hides, manure, and domestic fuel for the forming

Table 3.3: Characteristic Features of Small-Scale Dairy Management Systems [5]

System Type	*Extensive*	*Low Intensive*	*Moderately Intensive*	*Highly Intensive*	*Zero Grazing*
System No.	*I*	*II*	*III*	*IV*	*V*
Breeds and breeding	Local breeds	Cross cow (F_1)	Low upgraded cow	High grade dairy	High grade dairy
	Use of bull in natural service	Use of AI or bulls natural service	Use of AI		
	Uncontrolled mating	Controlled mating			
Rearing methods	Male and female calves suckle dams during lactation	Male and female calves part-time suckle a restricted amount of milk over a 4-7 month period	Bucket or hand rearing, feeding of whole milk, late weaning, 3-4 months	Rearing female calves only	Rearing female calves only
	Cow cannot be milked without a calf	It is difficult to milk the cow without a calf	Cow can be milked without having a calf	Concentrate	Bucket feeding
				Only weaning 2-3 months	Use milk substitutes
				Bucket feeding	Concentrate early weaning

Contd...

Table 3.3–Contd...

System Type	*Extensive*	*Low Intensive*	*Moderately Intensive*	*Highly Intensive*	*Zero Grazing*
System No.	*I*	*II*	*III*	*IV*	*V*
Forage production and feeding methods	Communal grazing of natural grass and bush land	Individual grazing of naturally regenerated pasture in a fallow system	Individual grazing on fenced and cultivated or improved pasture	Grazing cultivated land	Growing arable fodder
			Use of fertilizer	Feeding of arable fodder crops	Forage is cut and carried to animals
				Use of fertilizer	Fertilizer and manure are used
Cow feeding and management	Grazing during day and enclosing during night	Grazing during day and enclosing during night	Grazing day and night	Partially grazing	Stall feeding
	Feeding salt minerals	Feeding minerals	Feeding minerals	Feeding arable fodder crops in confined areas	Cattle kept permanently indoors

Contd...

Table 3.3–Contd...

System Type	*Extensive*	*Low Intensive*	*Moderately Intensive*	*Highly Intensive*	*Zero Grazing*
System No.	*I*	*II*	*III*	*IV*	*V*
		Feeding concentrates occasionally	Supplementary feeding concentrates	Feeding minerals	Feeding minerals
				Supplementary feeding of concentrates	Feeding concentrates regularly
Disease control	Rinderpest and foot-and-mouth	Tick-borne disease control	Tick-borne disease control		– do –
		Rinderpest and foot-and-mouth vaccinations	Mastitis treatment		
			Rinderpest and foot-and-mouth vaccinations		
Possible cash cropping enterprises	Cotton	Maize, coffee, tea and pyrethrum	Maize, coffee, tea and pyrethrum		– do –

*: A cow unit (CU) is defined as a cow plus followers.

community. The crop husbandry which has got comparatively more attention by the policy makers, scientists, and economists so far is somewhat seems to be reached a stage of saturation whereas the livestock sector is still emerging and thus, offers a great hope for booming Indian economy. This necessitates more attention to this sector in the years to come.

Livestock keeping in India and similar other countries has multiple objectives and dimensions. The play multiple roles in rural systems and economy and have a strong human dimension, as manifested through socio-cultural link and involvement of women. Besides their well-established role in agriculture livestock have crucial role in food security and as risk aversion mechanism for sustaining family, whenever there is crop failure. Role of livestock in generating employment and income in rural areas is well established and livestock development has become an important component of rural development programs *i.e.*, "Equity and extending benefits directly to women" can be achieved through livestock development, since livestock distribution is less skewed than land. Livestock is a part of nature's chain for recycling nutrients, converting low quality and other agro bye-products into good quality and organic fertilizer. The latter being important for retaining soil fertility and productivity in ecologically fragile hill region. Moreover the farmers always take holistic view and are good example of systems manager who has to make decision on variety of factors. The food demands and requirement are given in Table 3.4. The most important and efficient cropping system are given in Table 3.5.

Table 3.4: Food Demand and Requirements [6]

Commodity	*Demand Growth Rate (1995-2020)*	*Output Growth Rate in Last Ten Year*
Cereals	1.88	2.16
Pulse	2.99	0.63
Edible oil	2.91	2.06
Milk	3.26	4.14
Fruit	3.20	5.75
Vegetables	2.91	4.79
Eggs	3.76	4.59
Fish	3.75	4.28

Table 3.5: Efficient Cropping Systems [7]

State	*Crop Sequence*	
	Irrigated	*Rainfed*
Bihar	☆ Maize-Mustard/Yellow Sarson	☆ Maize/Early rice-Toria/Mustard/Yellow Sarson
	☆ Maize-Toria-Wheat-Moong	
Gujarat	☆ Bajra/Groundnut/Sesame-Mustard	—
	☆ Moong/Urd-Mustard	
Haryana	☆ Early fodder-Mustard	☆ Maize/Bajra-Mustard
	☆ Groundnut/Bajra-Mustard	
Madhya Pradesh	☆ Soybean (JS-90-41, JS-93-05, KB-79, LSB-1)/Fallow/Moong-Mustard	☆ Soybean-Mustard
Orissa	☆ Early rice-Toria-Summer sesame	☆ Early Rice-Toria/yellow sarson
	☆ Jute-Toria/Mustard-Summer sesame	
	☆ Rice-Toria-Rice	
Punjab	☆ Rice-mustard/gobhi sarson	☆ Bajra-Mustard
Rajasthan	☆ Maize/Bajra/Moong/cluster bean/ Cowpea-Mustard/Taramira	☆ Sorghum (Fodder)- Mustard
	☆ Bajra + Urd bean-Mustard	☆ Bajra-Mustard
		☆ Urd/Moong/Cowpea-Mustard

Contd...

Table 3.5–Contd...

State	Crop Sequence	
	Irrigated	*Rainfed*
Uttar Pradesh	☆ Maize-Autumn Sugarcane + Mustard	☆ Maize-Mustard
	☆ Maize-Toria-Wheat	☆ Maize/Bajra-Toria/Mustard/yellow sarson
	☆ Rice-Mustard/yellow sarson	☆ Rice-Mustard/yellow sarson
	☆ Sesame-Mustard	
	☆ Urd/Moong/Cowpea-Mustard	
Jammu & Kashmir	☆ Rice-Brown sarson	☆ Maize-Brown sarson
	☆ Rice-Oat	☆ Maize-Oat
	☆ Rice-Wheat	☆ Maize-Pea
	☆ Maize-Brown sarson	☆ Maize-Lentil
	☆ Maize-Oat	☆ Beans-Oat
	☆ Maize-Pea	☆ Beans-Brown sarson
		☆ Sunflower-Oat/Brown sarson
		☆ Mung-Oat/Brown sarson

3.4 Future Policy Review Point for Production Based Livelihood

- ☆ Ensuring peoples participation.
- ☆ Improving farm production and diversifying farming systems.
- ☆ Land-resource planning information and education for agriculture.
- ☆ Land conservation and rehabilitation.
- ☆ Water for sustainable food production and sustainable rural development.
- ☆ Conservation and sustainable utilization of plant genetic resources.
- ☆ Conservation and sustainable utilization of animal genetic resources.
- ☆ Integrated pest management control in agriculture.
- ☆ Sustainable plant nutrition to increase food production.
- ☆ Rural energy transition to enhance productivity.
- ☆ Evaluation of UV radiation damage resulting from depletion of the ozone layer.
- ☆ Investment of credit in agricultural enterprises must be increased.
- ☆ Specified market should be created.
- ☆ Recent agricultural commodity price information should be reached to farming community.

3.5 Key References and Web Resources

1. Hindu survey of Indian agricultures
2. Perfumer & Flavorist, 2005, 30(7), p.33
3. Economic survey of India (Gov. of India)
4. http://www.dorabjitatatrust.org/NGO_Grants/nrm_rural_live.aspx
5. http://www.ilri.org/html/trainingMat/policy_X5547e/x5547e26.htm
6. http://www.fao.org/docrep/x0172e/x0172e00.HTM
7. http://www.orissadiary.com/ShowBussinessNews.asp?id=11360

Chapter 4
Urban Wastes Management in Agriculture

The population growth has put tremendous pressure on the quality of Environment of urban life. The residents generate various kinds of wastes of biodegradable and non biodegradable categories. The impact created by these wastes on the environment is enormous, if proper disposal and management options are not applied. The waste could become a resource of bulk organic nutrients and the society can benefit from these wastes with proper collection and disposal technologies. Majority of wastes can be recycled and the recycling technologies available today world wide has a promising employment and energy generating options. The sanitary landfills, biogas generating technologies, vermicomposting, incineration, municipal solid waste combustion technologies offer good incentives for the local bodies and the governments to derive benefits by utilizing the bulk source of nutrients.

Keywords: *Urban waste, Municipal solid wastes, Industrial wastes, Pollution.*

The urban population has grown up at a rate of 19.0 per cent–34.6 per cent during last decades whereas the growth rate of total population is about 2 per cent. The cities of Bombay, Kolkata, Delhi and Chennai together accommodate 37.22 million population. One of the main reason for increase in the urbanization is the migration of rural population into cities for the employment. Each of the urban residents generates 350–1000 g solid wastes everyday. Taking waste from the commercial and industrial establishments the total waste generation/day adds up to an enormous quantity. The quantity of municipal solid wastes in a few important cities is given in Table 4.1.

Table 4.1: The Quantity of Municipal Solid Wastes in a Few Important Cities [1]

City	Generated Tonnes/day	Cleared Tonnes/day	Per cent Collected
Ahmedabad	1500	1200	80
Bangalore	2130	1800	85
Mumbai	5800	5000	86
Kolkata	3500	3150	90
Delhi	3880	2420	62
Lucknow	1500	1000	67
Chennai	2675	2140	80
Patna	1000	300	30
Surat	1250	1000	80

The public concern about pollution and its influence on ecology has placed great emphasis on the disposal of urban wastes. Deteriorating water quality makes it evident that basic practices of disposing of wastes in out streams can no longer be tolerated. Likewise the increase in smoke has led to added restriction on burning of refuse. In our search for alternative solutions to waste management problems more and more attention is being given to disposal of domestic and industrial wastes on land. The various categories of wastes generated in urban areas is given in Table 4.2.

It has long been article of faith among city dwellers that almost any form of waste with proper composting and processing can be made into a fertilizer that farmers will gladly pay for. They realize that even if it does not contain much of the essential plant nutrients, surely it will improve soil structure and produce healthy plants. This simply is not true enough to persuade the modern commercial farmer. The present day farm manager is a perfect businessman and he will first ask "what will it cost me to put this on the land per kg of available

Table 4.2: Various Categories of Wastes Generated in Urban Areas [1]

Catagories	*Source of Generation*	*Types of Solid Wastes*
Municipal solid wastes	*Residential:* Family dwelling, low, medium, and high rise buildings/Apartments	Food wastes, Rubbish Ashes, Special Wastes
	Commercial: Stores, Restaurants, Markets, Hotels, Shopping complexes, Repairshops etc.	Food wastes, Rubbish Ashes, Construction wastes, Special wastes, Occasionally hazardous wastes
	Open areas: Streets, Alleys, Parks, Vacant plots, Play grounds, Beaches, Highways, Recreational areas etc.	Special wastes and Rubbish
	Treatment plant sites: Water, Waste water and Industrial treatment processes	Treated wastes/Residual sludge.
Industrial wastes	Textiles, Paper and allied products, Chemicals and related products, Rubber, food kindred products, glass, petroleum and refineries etc.	Chemicals, Metals, Scrap products, Gypsum, Asbestos, Resins, Glass, Organic dyes, Glues etc.
Hazardous wastes	*Indusries:* Chemical and oil refineries, Ordinance factories, Fire works etc.	Volatile organic chemicals, Inflammable substances, Toxic gasses and Liquids etc.
	Hospitals: Bio-medical wastes	Chemicals, Pathological wastes, Infectious wastes, Sharp objects, Pharmaceutical wastes, Laboratory wastes, Pressurised canes etc.

N, P and K" and "Does it contain substances that will harm my soil or reduce the value of my crop". So in approaching the question of beneficial uses of urban wastes on land for crop production. It must be realized that benefits to the farmer are not to be minimal, on the other hand, benefits to those having to dispose of the wastes are obvious. This is not to say that agriculture has no role or responsibility in helping to handle the nations avalanche of wastes. But when the wastes are to be used on productive land, the economics will have to be understood and more importantly sustained productivity of the land and the quality of the drainage water from it must not been impaired.

The composition of municipal refuse is difficult to define because it may contain any and all substances used by man. Available information indicates that paper products make up one half the waste of urban waste, food waste make of one fifth and another one fifth is made up of such non degradable materials like metals, glasses, ashes etc. The characteristics of waste depends on various factors like food habits, cultural traditions of the inhabitants, life styles, climate etc. Constituents (per cent) of municipal solid wastes is mentioned in Table 4.3.

Table 4.3: Constituents (per cent) of Municipal Solid Wastes [1]

Components	*Composition (per cent)*
Vegetables/Leaves	40.15
Paper	3.8
Plastic	0.81
Leather/Rubber	0.62
Glass/Ceramic	0.44
Metal	0.64
Stone/Ashes/Dust	41.81
Miscellaneous	11.73

4.1 The Urban Waste Challenge

The accelerated growth of the global urban population implies an increasing demand for public services. Yet, urban centers in developing countries are unable to meet such demand. Services such as sanitation are poor or inadequate to cope with the increasing

Plate 4.1: Unmanaged Urban Waste

rates of urbanization and the associated higher standards of living. According to the UN 2002 Human Development Report, 2.4 billion people in the developing world lack access to basic sanitation. Indian cities and towns generate about 80,000 metric tonnes of municipal solid wastes everyday and it is estimated that about 25 million tonnes of such wastes is generated annually. However, on an average, only 60 per cent of solid wastes is collected in urban areas leaving the balance 40 per cent of the urban wastes unattended to. This gives rise to the unsanitary conditions and diseases, especially among the urban poor who constitute 40 per cent of urban population. Waste is a product or material that does not have a value anymore for the first user and is therefore thrown away; however, it could have value for another person in a different circumstance or even in a different culture [2]. As much as 90 per cent of the Municipal Solid Waste (MSW) collected in Asian cities end up in open dumps. The failure of city authorities to collect waste leads to unpleasant conditions and decomposing wastes constitute a serious health and environmental hazard. Urban waste could be solid or liquid, organic or inorganic, recyclable or non-recyclable. A considerable quantity of urban waste is biodegradable and hence is of immediate interest in recycling. The unmanaged urban waste show in Plate 4.1 and

Figure 4.1 shown the biosolid and liquid management using biosenitisor.

4.2 Recycling of Urban Organic Waste for Urban Agriculture

India's Green Revolution rescued the nation from famines, but left over 11.6 million hectares of low-productivity, nutrient-depleted soils ruined by unbalanced and excessive use of synthetic fertilizers and lack of organic manure or micronutrients. City compost can fill this need and solve both the problems of barren land and organic nutrient shortages, estimated at six million tons a year [1]. India's 35 largest cities alone can provide 5.7 million tonnes a year of organic manure if their biodegradable waste is composted and returned to the soil. Integrated plant nutrient management, using city compost along with synthetic fertilizers, can generate enormous national savings as well as cleaning urban India. Current urban organic waste recycling practices include the following: The use of fresh waste from vegetable markets, restaurants and hotels, as well as food processing industries as feed for urban livestock. Direct application of solid waste on and into the soil; Mining of old waste dumps for application as fertilizer on farm land [3]. Application of animal manure such as poultry/pig manure and cow dung; Direct application of human excreta or bio-solids to the soil [4]. Organized composting of SW or co-composting of SW with animal manure or human excreta. Whichever method is used, a process of microbial degradation releases the useful nutrients in organic waste for soil improvement and plant growth.

Sustainable management of solid waste is a major challenge being faced by municipal authorities across the world, both in the North and the South. Much of the solid waste consists of organic matter that can be recycled into a profitable input (compost) for urban agriculture. Composting the large quantities of organic matter provides a win-win strategy by reducing waste flows, enhancing soil properties, recycling valuable soil nutrients and creating livelihoods. Crop and hydrology model and optimisation tool for site specific management shown in Figure 4.2.

4.3 Benefits and Constraints

Potential benefits of organic waste recycling are particularly in reducing the environmental impact of disposal sites, in extending

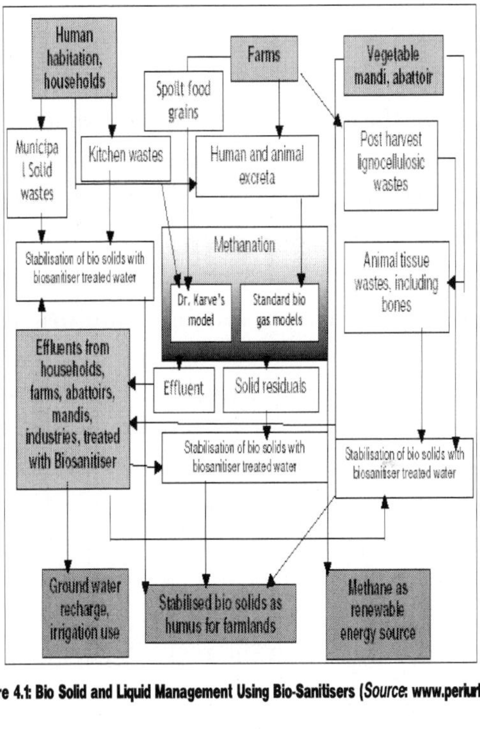

Figure 4.1: Bio Solid and Liquid Management Using Bio-Sanitisers (*Source*: www.periurban.org)

existing landfill capacity, in replenishing the soil humus layer and in minimizing waste quantity. Other benefits of composting are:

- ☆ Increases overall waste diversion from final disposal, especially since as much as 80 per cent of the waste stream in low- and middle-income countries can be composted.
- ☆ Enhances recycling and incineration operations by removing organic matter from the waste stream.
- ☆ Produces a valuable soil amendment–integral to sustainable agriculture.
- ☆ Promotes environmentally-sound practices, such as the reduction of methane generation at landfills.
- ☆ Enhances the effectiveness of fertilizer application.
- ☆ Can reduce waste transportation requirements.
- ☆ Is flexible for implementation at different levels, from household efforts to large scale centralized facilities.
- ☆ Can be started with very little capital and operating costs.
- ☆ The climate of many developing countries is optimum for composting.
- ☆ Addresses significant health impacts resulting from organic waste such as reducing Dengue Fever.
- ☆ Provides an excellent opportunity to improve a city's overall waste collection programme.
- ☆ Accommodates seasonal waste fluctuations such as leaf litter and crop residues.
- ☆ Can integrate existing informal sectors involved in the collection, separation and recycling of wastes.

Although composting seems an attractive option in many respects, it is also constrained by the following factors:

- ☆ Inadequate attention to the biological process requirements.
- ☆ Over-emphasis placed on mechanized processes rather than labour-intensive operations.
- ☆ Lack of vision and marketing plans for the final product–compost.

- Poor feed stock which yields poor quality finished compost, for example when contaminated by heavy metals.
- Poor accounting practices which neglect that the economics of composting rely on externalities, such as reduced soil erosion, water contamination, climate change, and avoided disposal costs.
- Difficulties in securing finances since the revenue generated from the sale of compost will rarely cover processing, transportation and application costs.

4.4 Composting & Vermiculture: The Environmentally and Economically Sustainable Solution from Urban Waste

It has been apparently referred to earthworms as the "intestines of the earth." Though composting of organic waste can take place without worms, a practice known as aerobic composting. The presence of these small creatures–particularly certain varieties–greatly expedites composting and improves the conversion of waste into nutrient-rich soil conditioner. However, there may be hazards associated with worm composting and the added expense may not make the resulting product commercially viable. The development and implementation of both types of composting is well documented in India, and the use of composting has attracted a great deal of attention from those outside the South Asia. Composting of city wastes is a legal requirement provided under the Municipal Solid Waste Management (MSW) Rules 2000 for all municipal bodies in the country. But neither the central nor the state governments have yet responded to show any kind of preparedness for it, nor have they been able to grasp it as an environmental and social good that requires official support which can generate employment. The MSW Rules 2000 requires that "biodegradable wastes shall be processed by composting, vermi-composting, anaerobic digestion or any other appropriate biological processing for the stabilization of wastes". By far, the better composting options are those that are decentralized and use organic waste as close to the source as possible. Decentralized on-site (for commercial organic waste) and on-plot (for domestic organic waste) are the preferred levels of intervention with each individual intervention requiring the appropriate technology at the appropriate scale. In essence, the primary function is all about getting the nutrients and organic matter in waste back into the soil in the most efficient and effective manner; hence the priority order of backyard composting (household) and decentralized (community) approaches. Centralized municipal approaches do not have a good track record and the potential scale-of-economy advantages have not materialized due to operational and marketing constraints.

4.5 Ecofriendly Management

A very attractive way to change garbage into rich humus is to utilize the services of earthworms. Vermiculture means farming of earthworms through bio-degradable material. Earthworms are nature's fertilizer factory. Physically they are crushers and grinders, due to action of their gizzard. There are thousands of different species of worms, but the best manure worms is *Eisenia foetida,* as it works everywhere, in the indoor as well as at outdoor. They are a surface dwelling variety of worms that hate the light and reproduce at an amazing rate. In urban areas for treatment and conversion of household waste into high quality compost, a package has been developed by Morarka Foundation.

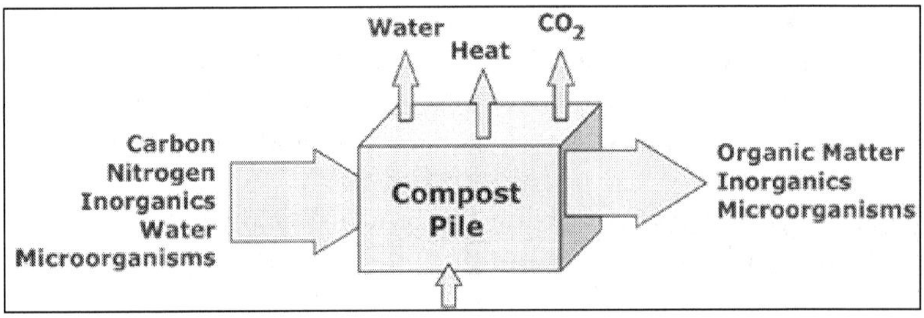

The package aims at:

- Creating awareness at household level regarding the issue of garbage and its proper management.
- Collection of wastes in segregated form.
- Conversions of organic/wet waste into high quality vermicompost *i.e.* put your waste in and get the vermicompost out there by recycling the nutrients.
- Appropriate use of vermicompost for planting trees, gardens, lawns, etc. to make clean and green environment surrounding our houses.

4.6 Scientific Waste Management

In urban areas, waste including human excreta and waste from polluting industries are disposed through sewers. This pollutes the environment, under ground water and exposes people to infection. Open decomposition of solid waste and sewer water through existing river systems takes very long period in natural treatment while causing many health hazards. In many countries, initiatives to develop the skills through recycled waste materials for producing vegetables by the poor have brought about excellent results. In some Latin American countries, vegetable production in urban areas through waste recycling have not only been able to reduce the direct and indirect costs associated with waste disposal but it has also simultaneously been able to solve the problems of urban sanitation, while becoming an income generating activity as well. In view of above, a new approach has therefore been developed by Morarka Foundation to utilize biotechnologies for recycling of waste materials. Production of foods from recycled wastes is now also known as eco-sanitation.

4.6.1 Use of Industrial and Urban Wastes in Agriculture

- ☆ Real time monitoring of soil, air and water pollution caused by wastes, especially of distilleries, paper mills and sewage treatment plants. The Plate 4.2 showed wasteater application..

Plate 4.2: Wastewater Applications in Agro-ecosystems

- ☆ Feasibility options of utilization of wastes in agriculture and predicting long-term consequences of wastewater applications in agro-ecosystems through simulation techniques.
- ☆ Development of protocols/policy for sustainable use of effluents, and their monitoring.

4.6.2 Solid Wastes Management

- ☆ Biogas production from agri- residues and wastes.
- ☆ Solid State fermentation for the production of energy and manure.
- ☆ Manurial potential of municipal solid wastes, composting, and its utilization in agriculture (Plate 4.3).

Plate 4.3: Municipal Solid Wastes, Composting, and its Utilization in Agriculture

4.6.3 Air Pollution

- ☆ Impact of fly ash on agricultural crop production and soil properties.
- ☆ Utilization of fly ash for soil amendment.
- ☆ Monitoring the status of heavy metals in soils, water, and crops due to air pollution.

4.6.4 Environmental Indicators

- ☆ Development of comprehensive indicators of environmental status of soil, air and water in agro-ecosystems.
- ☆ Rating of different production systems/regions based on environmental indicators, especially resource use efficiency.
- ☆ Identification of early warning signals of stress in agro-ecosystems.

4.6.5 Crop and Soil Ecology

- ☆ Impact of pesticides on soil biota.
- ☆ Impact of resource conservation technologies on soil microbial biodiversity.
- ☆ Crop-pest ecology.
- ☆ Modelling crop response to climatic, edaphic, biotic and production variables.
- ☆ Gene escape through rhizosphere of genetically modified crop plants.

4.6.6 Sustainable Land Use Systems

- ☆ Using GIS, remote sensing, participatory rural appraisals, surveys, crop and hydrology models, and optimization tools for site-specific management (Figure 4.2).
- ☆ Decision Support Systems (DSS) integrating natural resources inventory, simulation of soil, water and crop processes, and socio-economics for optimal land use and conservation practices in watersheds.
- ☆ Field evaluation of DSS and its recommendations.

4.6.7 Biofuels

- ☆ Genetic enhancement of crops for higher starch/sugar content for their utilization in ethanol production.
- ☆ Production technology for ethanol using grain and biomass of crops (Plate 4.4).

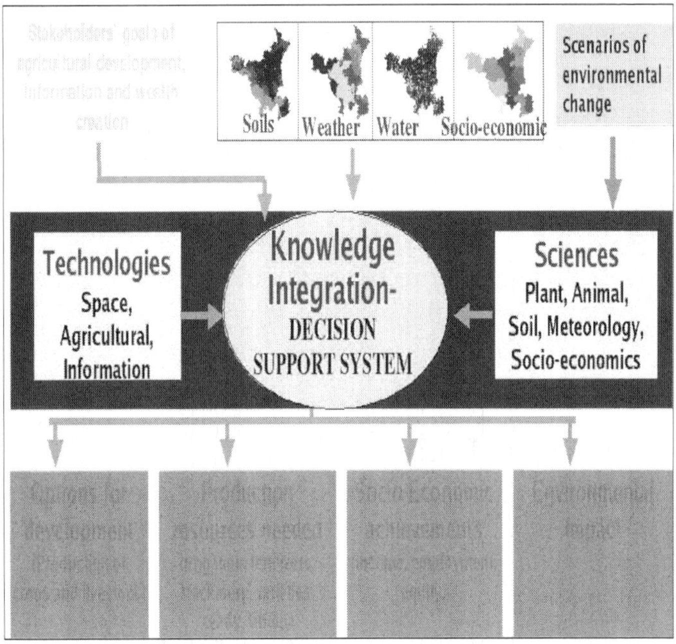

Figure 4.2: Crop and Hydrology Model and Optimization Tools for Site-specific Management

Plate 4.4: Jatropha Plantation in Urban Agricultural Fields

- ☆ Evaluation of non-edible oilseeds, especially Jatropha, for biodiesel production.
- ☆ Production technology of Jatropha in urban agricultural fields (Plate 4.4).
- ☆ Utilization of by-products of ethanol and biodiesel production processes in urban agriculture.
- ☆ Production economics of using agricultural resources for biofuel.

4.6.8 Bio-medical Composting

This process of bio-culture cultivation has proved useful in the safe disposal of organic hospital waste, including human placentas, sanitary napkins, and all kinds of dressing material soaked in blood and serum. For this construct a cement pit of 6 by 6 feet, with two compartments. In the bottom compartment, specific bio-culture has been scattered, and in here goes all the garbage that has been segregated. Once an adequate amount of decomposed manure has been prepared from this waste, it is removed from the top

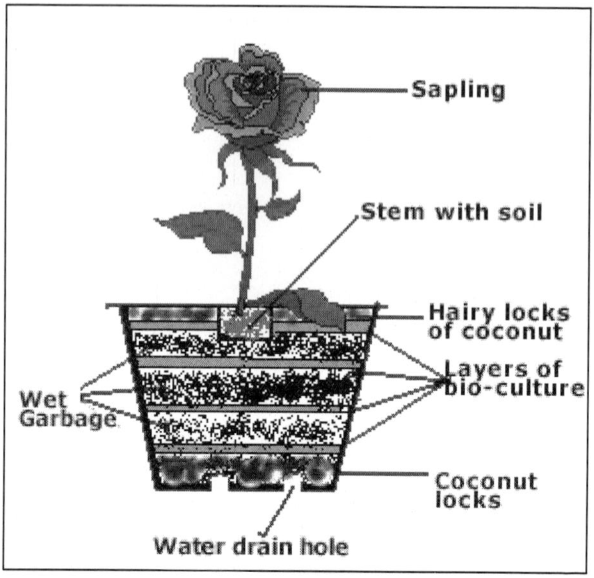

Plate 4.5: Hospital Waste Decomposed Bio-culture Used in the Neighbouring Garden

compartment to be used in the neighbouring garden. "There is no risk, and compost can be removed even with bare hands." This method is definitely quicker, and safer, compared to earlier practice of burning the waste. The Plate 4.5 shows the utilization of hospital waste decomposed bio-culture in neighbouring garden.

4.6.9 Wastewater Use in Urban and Peri-urban Agriculture and its Contribution to Food Security

Urban and peri-urban farmers from different caste and class groups in developing countries in Asia and Africa derive their livelihoods by using wastewater for various activities such as horticulture, fodder production for dairy activities, agroforestry, orchard keeping, floriculture, aquaculture and cereal production. There are also many areas in which the government runs sewage farms near treatment plants which are hired out to farmers for cultivation such as those around Madurai, South India. Wastewater users, who come from a wide range of socio-economic backgrounds, have a variety of motives for using wastewater for irrigation. In semi-arid and arid areas it is often the only source of water available in sufficient quantities for irrigation; it is also available year-round unlike freshwater from rainfall which is concentrated in the often short and sporadic rainy season. It is also an inexpensive source, not only of water but also of nutrients. In fact, farmers often need few or no additional fertilizers. Crop yields are often higher with wastewater than with freshwater.

4.7 Restoring Degraded Soils in India using Urban Wastes

Coal-fired turbines generate four-fifths of India's electricity. The 200 million tonnes of low-grade coal that they burn each year discharge up to 100 million tonnes of fly ash into enormous settling ponds, causing siltation, flooding, and contamination of water sources for millions of people [5]. As well, every city in India produces huge amounts of mostly untreated sewage sludge causing water contamination risk. Finally, water hyacinth, a free floating weed introduced from South America before 1900, now infests an estimated 200,000 hectares of Indian waterways, choking off plant, fish, and animal growth.

Yet mixing the three materials creates a tonic, rather than a toxic, for soils where nothing has grown for a century or more, with some primary treatment, sewage sludge can be mixed with fly ash and sometimes hyacinth to yield more than just a fertilizer. Combining the sludge's nitrogen and organic matter with the minerals found in fly ash yields a potent soil replacement substance. Fly ash does a lot for the soil [6]. It reduces bulk density, increases water holding capacity, buffers pH (soil acidity), and adds both macro and micro nutrients. The major elements are potassium, phosphorus, calcium, magnesium, and carbon from unbound coal. Potential trace elements include boron, molybdenum, selenium, nickel, copper, zinc, and many exotic elements whose functions are not fully understood in plant physiology. The trick is getting just enough to be beneficial, but not enough to be toxic.

4.8 Applications

So far, the data show that vegetation grown on lands treated with this mix absorbs low levels of heavy metals. For now, the Indian private sector is using it to grow commercial tree species for plywood, and some sugar cane. Public-sector users have hopes to produce non-timber forest products for local villages, such as fuel wood, animal feed, medicinal plants, and grasses on marginal/wasteland and salt effected soils. The researchers have also planted small plots of edible crops to compare different ratios of ash and sludge, and to analyze the plants for metal uptake from the reconditioned soil.

All the data to date prove that the amount of metal uptake is within international guidelines, says. There are a couple of exceptions in the case of lead and chromium, but even these aren't far from the upper allowable limits. This technology can be used on edible crops, but caution that we are still in the experimental phases.

4.9 Concentration Not to Exceed

Ministry of Environment & Forests vide gazette notification S.O.908 (E) dated 03.10.2000 has set guidelines and maximum permissible limits of heavy metals and impurities as per following table, which need to be followed for selling urban compost for production of food crops. The maximum permissible limits of heavy metals are given in Table 4.4.

Table 4.4: Maximum Permissible Limits of Heavy Metals and Impurities in Urban Waste for Agriculture Use (mg/kg dry basis, except pH value and C/N ratio)

Arsenic	10.00	Cadmium	5.00
Chromium	50.00	Copper	300.00
Lead	100.00	Mercury	0.15
Nickel	50.00	Zinc	1000.00
C/N ratio	20/40	pH	5.5–8.5

4.10 Neutralizing Industrial Waste with Worms

Vermi-composting to convert household waste into manure is widely used worldwide, but using it to treat toxic waste is relatively recent and yet to gain acceptance [7]. While the sludge settles down, the clearer water outflows into outer drainage system. It is this sludge which still has about 70-80 per cent water and can be dewatered by a belt press filter. The remaining 20-30 per cent solid sludge contains organic as well as inorganic material. Depending upon the characterization of the sludge, a predetermined proportion is spread on a vermi-culture bed that uses earthworms. Like in conventional vermin-composting, the organic matter from the sludge gets converted into manure.

But what happens to the toxic materials such as heavy metal compounds in the sludge? These can kill earthworms. Yes, scientists, confirming that the toxic material gets adsorbed by the earthworms[7]. However, his research shows that while the average life of earthworms is one and half year, depending upon the concentration of the toxic matter, it reduces to 6-8 months. Even within this shortened span of life, earthworms regenerate 10-20 times. Dead worms too form organic matter. The same quantity of toxic matter gets distributed in multiplied worms leaving manure toxic free [7].

4.11 Salient Achievements

- ☆ Many of the industrial effluents contain considerable amounts of plant nutrients and can be used in agriculture as a source of plant nutrient. Our research has shown that post methanation distillery effluent (PME) and paper mill effluent can be used in agriculture as sources of plant nutrients. Application of PME either pre-sown or post sown increased yields of rice, wheat, mustard, sugarcane and medicinal plants like *Mentha arvensis* significantly compared to recommended levels of N, P and K application. Apart from increasing crop yields application of PME was effective in improving soil organic carbon, N, P, K and S status of soil. PME also has a good potential to ameliorate the sodic soils. The Division has developed a protocol for use of PME, which has been accepted by the Ministry of Environment and Forests, Government of India to be implemented in all the distilleries in India.
- ☆ Disposal of fly ash produced at the thermal power plants is a great problem since it is not used in large scale in other sectors. The Division of Environmental Sciences showed that there were no adverse effects on crop yields with ash incorporation and safe limits of fly ash use were prescribed. Presently the division is working on the possibility of using municipal solid waste in agriculture.
- ☆ The Division has studied the air pollution impact on vegetables in Delhi. The vegetable samples collected from different locations in Delhi manifested higher level of heavy metal (Zn, Cu, Pb and Cd) contamination. Washing of vegetable samples twice and thrice reduced the level of

heavy metal contamination drastically contributed through air pollution.

- ☆ Several computer based Decision Support Systems (DSS) have been developed to assist in environmental impact assessment.
- ☆ Identified and optimized the process conditions for maximizing the biogas production from different agricultural wastes such as crop residues, fruit and vegetable wastes, and aquatic weeds. Developed Dry Fermentation Technology (Solid State fermentation) for the production of energy and manure from agricultural residues and kitchen wastes. Unlike the conventional biogas plant, the technology can accommodate all types of fibrous organic wastes as alternate and supplemental feedstock to cow dung for the production of biogas.

4.12 References

1. Upadhyay, V.P, Prasad, M. R., Srivastav, A. and Singh, K. (2005) Eco-tools for waste management in India. J. Hum. Ecol., 18(4): 253-269.

2. Anonymous (1990) National Compositing Programme. Manilla; Deptt. of DSE publications.

3. Smith, J. (1994) Urban Agriculture, Progress and Prospect, 1975-2005. The Urban Agriculture network (TUAN). March 1996, pp 38-40.

4. Nair, G. K. (2003) Urban waste remains a mounting problem. Business line. Financial daily from the Hindu group of publications. Tuesday, Oct. 21.

5. Bhatia, Arti, Pathak, H. and Joshi, H. C. (2001) Use of sewage as a Source of plant nutrient potential and problems. Fertilizer News 46 (3): 55-65.

6. Pasquini, M. W. and Alexander, M. J. 2004. Chemical properties of urban waste ash produced by open burning on the Jos plateau: Implications for Agriculture. Science of total envoironment Vol. 319: 225-240.

7. Surekha, S., and Dabke, S. (2008) Neutralizing industrial wastes with worms. conceptbiotech @yahoo.com

4.13 Other References Consulted

1. Arnon, 1995. Clean energy from municipal solid waste. A vital bio energy technology for 21st century. IEA Bioenergy Task. X1 VETSU, Harewell UK.
2. Morarka, M. R -GDC rural research foundation. www.morarkango. com.
3. Biosolid and liquid management using Bio- Sensitors. www.periurban.org.
4. Municipal solid waste (Management and Handling) Rules 2000, Ministry of Environment and forest, Govt. of India, issues on 25 Sept. 2000.

Chapter 5

Climate Change and Mitigating Strategies for Sustainable Crop Production

Now-a-days climate change has become one of the burning issues of debate as it has endangered the very existence of life on earth. The factors which are mainly responsible for climate change has been broadly due to (1) Geo-ecological events (earthquake, tsunami, global warming/cooling/diming and long terms changes in the earth's climate (2) man kind intervention (increased cropped area/intensity of cultivation, deforestation, water-storage/dams, incursion of sea water, increased use of ground water/irrigation, less than 30 per cent forests of the total land area of an country, more than 30 per cent cropped area compared to total geographical area, mono-cropping, unbalanced use of biocides and chemicals, decreased recharge of ground water). The devastating impact of these factors had resulted in increased environmental temperatures, more variable weather, lowering of ground water, contaminated soils and ground water, acid–rain, higher soil erosion/land degradation, variable crop yields, decreased factor productivity, more water deficits–frequent droughts, higher frequency of floods, lower/decreased soil quality, human unrest, increased migration, livelihood insecurity, food and water insecurity and decreased quality of life. To overcome the impact of climate change we should adopt land/water conservation agricultural methods suited to varied agro-climates, increase crop diversity by inter-cropping and appropriate cropping systems/rotations/land use. Balanced use of biocides and chemicals. Increase the forest area up to 33 per cent of the total geographical area, increase carbon fixation in the soil by growing deep-rooted crops so as to decrease carbon foot–print. Use water judiciously (more crop/unit of water). Use less fossil fuel by using more solar/wind sources of energy. Educate farmers on the dangers of climate change. Disseminate meteorological/climate data/information on a large–scale. Suggest weather-based changes in cropping systems/land uses to sustain agricultural production. Encourage farmer groups to establish small weather observatories in their villages. Adopt use of soil-health cards widely for making fertilizer use decisions. Employ crop-weather models dynamically to advice farmers on improved animal/crop management for sustainable agriculture in a Decision Support System's framework.

Keywords: *Climate change, earth quake, Tsunami, Global warming/cooling.*

5.1 Factors Responsible for Climate Change

5.1.1 Due to Geo-ecological Events

In a study first published on the web in 2004, NASA and United States Geological Survey (USGS) scientists found that retreating glaciers in southern Alaska may lead to more earthquakes in future. "The study examined the likelihood of increased earthquake activity in southern Alaska as a result of rapidly melting glaciers. As glaciers melt they lighten the load on the Earth's crust. Tectonic plates, that are mobile pieces of the Earth's crust, can then move more freely." The study appeared in the July 2004 issue of the Journal of Global and Planetary Change. *Source*: www.nasa.gov/centers/goddard/news/topstory/2004/0715glacierquakes.html

5.1.1.1 Earthquakes

An earthquake is the result of a sudden release of stored energy in the Earth's crust that creates seismic waves. Earthquakes are accordingly measured with a seismometer, commonly known as a seismograph. The magnitude of an earthquake is conventionally reported using the Richter scale or a related Moment scale (with magnitude 3 or lower earthquakes being hard to notice and magnitude 7 causing serious damage over large areas). At the Earth's surface, earthquakes may manifest themselves by a shaking or displacement of the ground. Sometimes, they cause tsunamis, which may lead to loss of life and destruction of property. An earthquake is caused by tectonic plates getting stuck and putting a strain on the ground. The strain becomes so great that rocks give way by breaking and sliding along fault planes. Earthquakes may occur naturally or as a result of human activities. Smaller earthquakes can also be caused by volcanic activity, landslides, mine blasts, and nuclear experiments. In its most generic sense, the word earthquake is used to describe any seismic event–whether a natural phenomenon or an event caused by humans.

For one of the "team of scientists" that reported on the Greenland earthquakes now think that the earthquakes were the result of processes involved with glacial calving, rather than something "underway deep within the second largest accumulation of ice on the planet" [1].

Earlier this year 2006, yet another team of scientists reported that the previous twelve months saw 32 glacial earthquakes on Greenland between 4.6 and 5.1 on the Richter scale–a disturbing sign that a massive destabilization may now be underway deep within the second largest accumulation of ice on the planet, enough

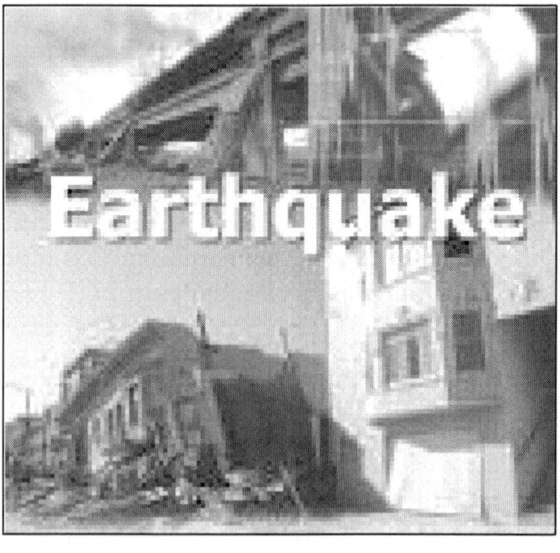

Plate 5.1: Building Collapse Due to Earthquake

Plate 5.2: Land Crakes Due to Earthquake

ice to raise sea level 20 feet worldwide if it broke up and slipped into the sea. Each passing day brings yet more evidence that we are now facing a planetary emergency–a climate crisis that demands immediate action to sharply reduce carbon dioxide emissions worldwide in order to turn down the earth's thermostat and avert catastrophe. Plate 5.1 shown collapse building and Plate 5.2 shown land crakes due to earthquakes.

5.1.1.2 The Tsunami's (Sea surface Temperature)

A tsunami is a series of waves created when a body of water, such as an ocean is rapidly displaced. Earthquakes, mass movements above or below water, volcanic eruptions and other underwater explosions, landslides, large meteorite impacts and testing with nuclear weapons at sea all have the potential to generate a tsunami. The effects of a tsunami can range from unnoticeable to devastating. The term tsunami comes from the Japanese words meaning *harbor*. For the plural; one can either follow ordinary English practice and or use an invariable plural as in Japanese. The term was created by fishermen who returned to port to find the area surrounding their harbor devastated, although they had not been aware of any wave in the open water. Tsunami's are common throughout Japanese history; approximately 195 events in Japan have been recorded. A tsunami has a much smaller amplitude (wave height) offshore, and a very long wavelength (often hundreds of kilometers long), which is why they generally pass unnoticed at sea, forming only a passing

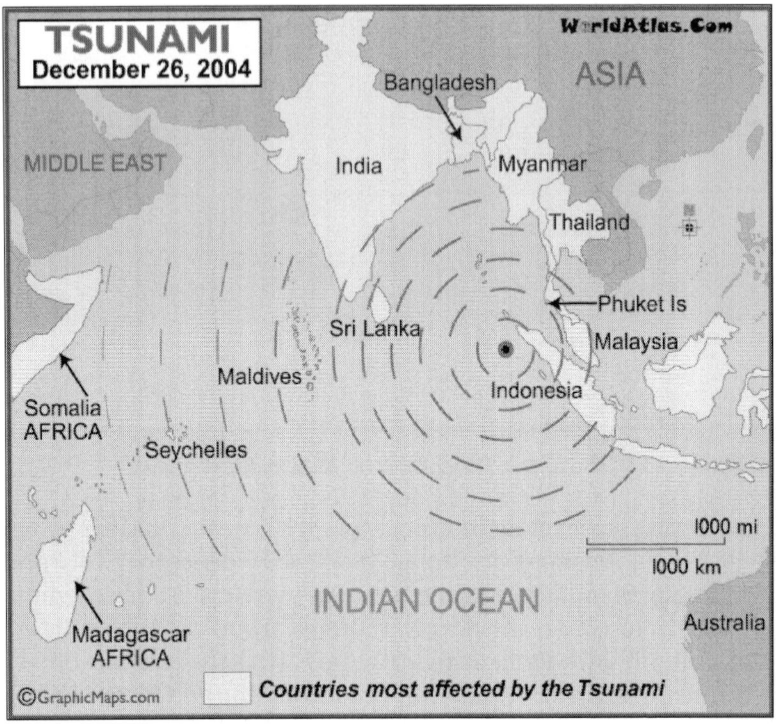

Plate 5.3: How Tsunami's Wave Forward

Plate 5.4: Disasters of Tsunami's

"hump" in the ocean. Tsunami's have been historically referred to *tidal waves* because as they approach land, they take on the characteristics of a violent on rushing tide rather than the sort of cresting waves that are formed by wind action upon the ocean (with which people are more familiar). Since they are not actually related to tides the term is considered misleading and its usage is discouraged by oceanographers (Plate 5.3).

We know that the planet's climate is changing. We know that we're largely responsible. We know that the poorest people on Earth will be hardest hit. We also know how to do many things to help them live better now, and better withstand the heavy weather that's on its way. The only question is whether we care enough to put our knowledge to use. (*Source*: http://www.worldchanging.com/archives/002906.html). Disasters of tsunamies shown in Plate 5.4.

5.1.1.3 Global Warming

Global Warming is defined as the increase of the average temperature on Earth. As the Earth is getting hotter, disasters like hurricanes, droughts and floods are getting more frequent. Over the last 100 years, the average temperature of the air near the Earth´s surface has risen a little less than 1° Celsius (0.74 ± 0.18°C, or 1.3 ± 0.32° Fahrenheit). Does not seem all that much? It is responsible for the conspicuous increase in storms, floods and raging forest fires we have seen in the last ten years, though, say scientists. Their data show that an increase of one degree Celsius makes the Earth warmer now than it has been for at least a thousand years. Out of the 20 warmest years on record, 19 have occurred since 1980. The three hottest years ever observed have all occurred in the last eight years, even.

The average facade temperature of the globe has augmented more than 1 degree Fahrenheit since 1900 and the speed of warming has been almost three folds the century long average since 1970. This increase in earth's average temperature is called Global warming. More or less all specialists studying the climate record of the earth have the same opinion now that human actions, mainly the discharge of green house gases from smokestacks, vehicles, and burning forests, are perhaps the leading power driving the fashion.

The gases append to the planet's normal greenhouse effect, permitting sunlight in, but stopping some of the ensuing heat from

Plate 5.5: Melting Glaciers

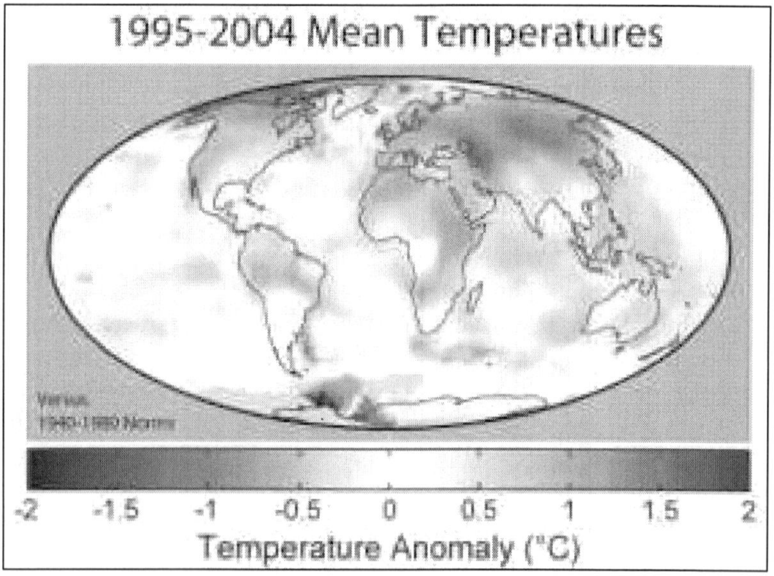

Plate 5.6: Mean Surface Temperature Anomalies during the Period 1995 to 2004 with Respect to the Average Temperatures from 1940 to 1980

radiating back to space. As said, the major cause of global warming is the emission of green house gases like carbon dioxide, methane, nitrous oxide etc. into the atmosphere. The major source of carbon dioxide is the power plants. These power plants emit large amounts of carbon dioxide produced from burning of fossil fuels for the purpose

Plate 5.7: Melting Glaciers on Himalayas
(*Source*: India today, 6 Nov 2006)

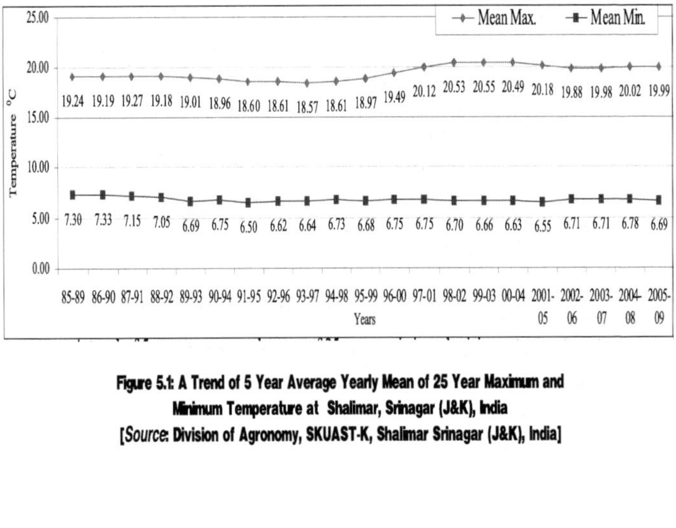

Figure 5.1: A Trend of 5 Year Average Yearly Mean of 25 Year Maximum and Minimum Temperature at Shalimar, Srinagar (J&K), India
[*Source*: Division of Agronomy, SKUAST-K, Shalimar Srinagar (J&K), India]

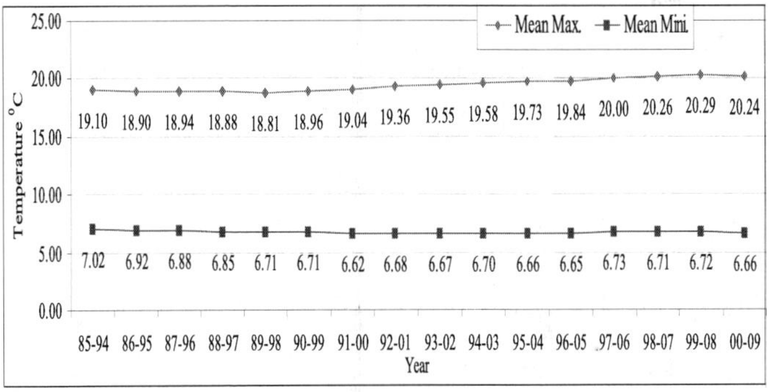

Figure 5.2: A Trend of 10 Year Average Yearly Mean of 25 Year Maximum and Minimum Temperature at Shalimar, Srinagar (J&K), India
[*Source*: Division of Agronomy, SKUAST-K, Shalimar Srinagar (J&K), India]

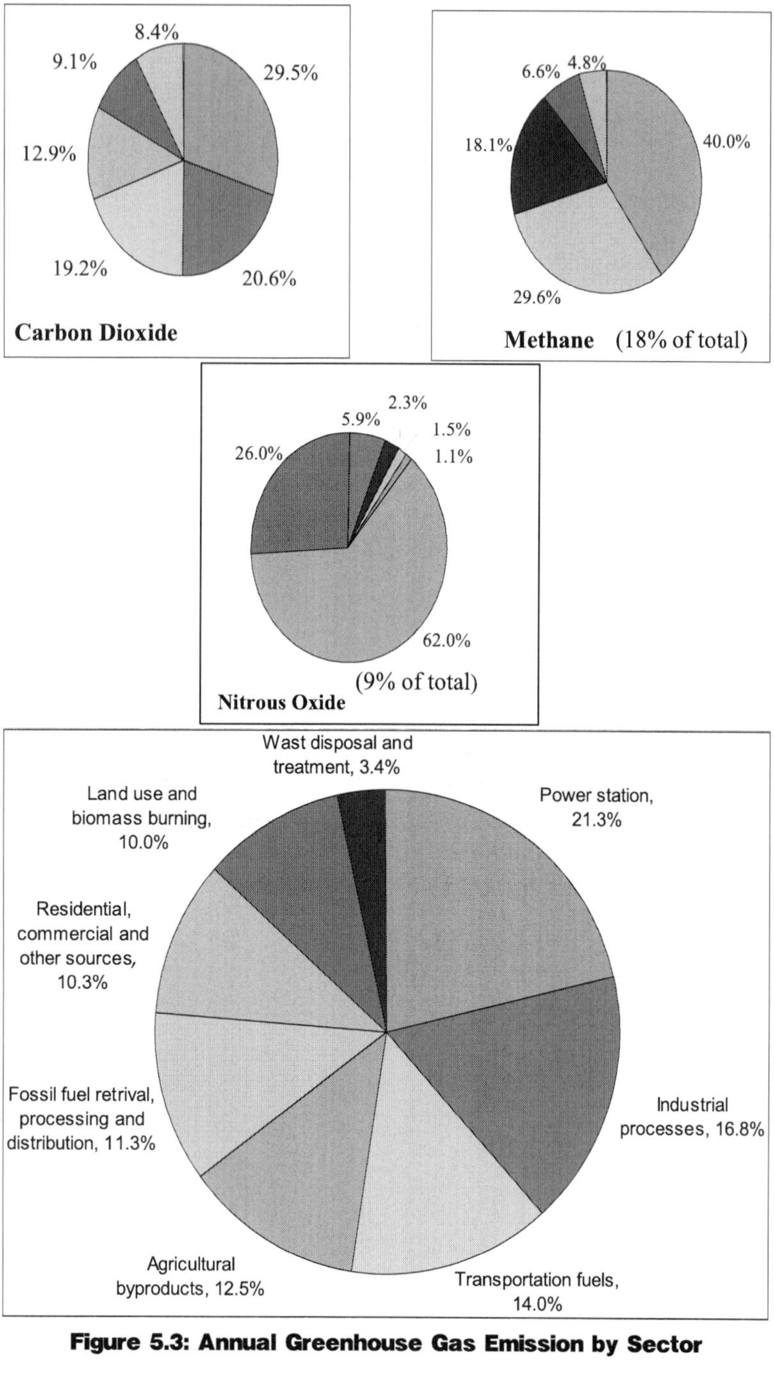

Figure 5.3: Annual Greenhouse Gas Emission by Sector

of electricity generation. About twenty percent of carbon dioxide emitted in the atmosphere comes from burning of gasoline in the engines of the vehicles. This is true for most of the developed countries. Buildings, both commercial and residential represent a larger source of global warming pollution than cars and trucks. Methane is more than 20 times as effectual as CO_2 at entrapping heat in the atmosphere. Methane is obtained from resources such as rice paddies, bovine flatulence, bacteria in bogs and fossil fuel manufacture. When fields are flooded, anaerobic situation build up and the organic matter in the soil decays, releasing methane to the atmosphere. The main sources of nitrous oxide include nylon and nitric acid production, cars with catalytic converters, the use of fertilizers in agriculture and the burning of organic matter. Another cause of global warming is deforestation that is caused by cutting and burning of forests for the purpose of residence and industrialization. The sector wise annual greenhouse gass emission shown in Figure 5.3.

5.1.1.4 Global Cooling

In the 1970s there was increasing awareness that estimates of global temperatures showed cooling since 1945. Of those scientific papers considering climate trends over the 21st century, only 10 per cent inclined towards future cooling, while most papers predicted future warming.[2] The general public had little awareness of carbon dioxide's effects on climate, but Science News in May 1959 forecast a 25 per cent increase in atmospheric carbon dioxide in the 150 years from 1850 to 2000, with a consequent warming trend.[3] The actual increase in this period was 29 per cent. Paul R. Ehrlich mentioned climate change from greenhouse gases in 1968.[4] By the time the idea of global cooling reached the public press in the mid-1970s temperatures had stopped falling, and there was concern in the climatological community about carbon dioxide's warming effects.[5] In response to such reports, the World Meteorological Organization issued a warning in June 1976 that a very significant warming of global climate was probable.[6] The cooling period is well reproduced by current (1999) on global climate models (GCMs) that include the physical effects of sulphate aerosols, and there is now general agreement that aerosol effects were the dominant cause of the mid-20th century cooling. However, at the time there were two physical mechanisms that were most frequently advanced to cause cooling (i) aerosols and (ii) orbital forcing.

Human activity–mostly as a by-product of fossil fuel combustion, partly by land use changes–increases the number of tiny particles (aerosols) in the atmosphere. These have a direct effect (they effectively increase the planetary albedo, thus cooling the planet by reducing the solar radiation reaching the surface) and an indirect effect (they affect the properties of clouds by acting as cloud condensation nuclei).[7] Orbital forcing refers to the slow, cyclical changes in the tilt of Earth's axis and shape of its orbit. These cycles alter the total amount of sunlight reaching the earth by a small amount and affect the timing and intensity of the seasons. This mechanism is believed to be responsible for the timing of the ice age cycles, and understanding of the mechanism was increasing rapidly in the mid-1970s. Greenhouse gases were regarded as likely factors that could promote global warming, while particulate pollution blocks sunlight and contributes to cooling.

5.1.1.5 Global Dimming

Global dimming is the gradual reduction in the amount of global direct irradiance at the Earth's surface that was observed for several decades after the start of systematic measurements in the 1950s. The effect varies by location, but worldwide it has been estimated to be of the order of a 4 per cent reduction over the three decades from 1960–1990. It is thought to have been caused by an increase in particulates such as sulfate aerosols in the atmosphere due to human action. The switch from a "global dimming" trend to a "brightening" trend in 1990 happened just as global aerosol levels started to decline. Global dimming has interfered with the hydrological cycle by reducing evaporation and may have reduced rainfall in some areas. Global dimming also creates a cooling effect that may have partially masked the effect of greenhouse gases on global warming.

It is thought that global dimming is probably due to the increased presence of aerosol particles in the atmosphere caused by human action.[8] Aerosols and other particulates absorb solar energy and reflect sunlight back into space. The pollutants can also become nuclei for cloud droplets. Water droplets in clouds coalesce around the particles.[9] Increased pollution causes more particulates and thereby creates clouds consisting of a greater number of smaller droplets (that is, the same amount of water is spread over more droplets). The smaller droplets make clouds more reflective, so that more incoming sunlight is reflected back into space and less reaches the Earth's surface. In models, these smaller droplets also decrease rainfall.[10] Clouds intercept both heat from the sun and heat radiated from the Earth. Their effects are complex and vary in time, location, and altitude. Usually during the daytime the interception of sunlight predominates, giving a cooling effect; however, at night the re-radiation of heat to the earth slows the earth's heat loss. The incomplete combustion of fossil fuels (such as diesel and wood) releases black carbon into the air. Though black carbon, most of which is soot, is an extremely small component of air pollution at land surface levels, the phenomenon has a significant heating effect on the atmosphere at altitudes above two kilometers (6,562 ft). Also, it dims the surface of the ocean by absorbing solar radiation.[11] Some research shows that black carbon will actually increase global warming, being second only to CO_2. They believe that soot will absorb solar energy and transport it to other areas such as the Himalayas where glacial melting occurs. It can also darken Arctic ice reducing reflectivity and increasing absorption of solar radiation.[12]

Large scale changes in weather patterns may also have been caused by global dimming. Climate modelers speculatively suggest that this reduction in solar radiation at the surface may have led to the failure of the monsoon in sub-Saharan Africa during the 1970s and 1980s, together with the associated famines such as the Sahel drought, caused by Northern hemisphere pollution cooling the Atlantic.[13] Because of this, the Tropical rain belt may not have risen to its northern latitudes, thus causing an absence of seasonal rains. This claim is not universally accepted and is very difficult to test. A natural form of large scale environmental shading/dimming has been identified that affected the 2006 northern hemisphere hurricane season. The NASA study found that several major dust storms in June and July in the Sahara desert sent dust drifting over the Atlantic Ocean and through several effects caused cooling of the waters–and thus dampening the development of hurricanes.[14][15] Some scientists now consider that the effects of global dimming have masked the effect of global warming to some extent and that resolving global dimming may therefore lead to increases in predictions of future temperature rise.[16] Global dimming interacts with global warming by blocking sunlight that would otherwise cause evaporation and the particulates bind to water droplets. Water vapor is the major greenhouse gas. On the other hand, global dimming is affected by evaporation and rain. Rain has the effect of clearing out polluted skies.

5.1.1.6 Long Terms Changes in the Earth's Climate

A rise in earth's temperatures can in turn root to other alterations in the ecology, including an increasing sea level and modifying the quantity and pattern of rainfall. These modifications may boost the occurrence and concentration of severe climate events, such as floods, famines, heat waves, tornados, and twisters. Other consequences may comprise of higher or lower agricultural outputs, glacier melting, lesser summer stream flows, genus extinctions and rise in the ranges of disease vectors. As an effect of global warming various new diseases have emerged lately. These diseases are occurring frequently due to the increase in earth's average temperature since the bacteria can survive better in elevated temperatures and even multiplies faster when the conditions are favorable. The global warming is extending the distribution of mosquitoes due to the increase in humidity levels and their frequent growth in warmer atmosphere. Various diseases due to ebola, hanta and machupo virus are expected due to warmer climates. A trend of change in temperature at shalimar are given in Figures 5.1 and 5.2.

5.1.1.7 Incursion of Sea Water

According to the Intergovernmental Panel on Climate Change, Bangladesh is slated to lose the largest amount of cultivated land globally due to rising sea levels. A 1m rise in sea levels would inundate 20 per cent of the country's landmass. "It is clear that climate change is taking its toll in the form of saline water intrusion into the mainland of Bangladesh, which is one of the lowest-altitude countries in the world". Most of the affected area is less than 1.5m above sea level. With every rising tide, sea water deposits salt on the land. Cultivation of rice, a food staple, has suffered most, while the production of wheat, pulses, rape seed and coconut has also been affected. And despite the fact that there is no official record of reduced agricultural output due to increased salinity in the soil, analysts say the drop could be as much as 50 per cent over the past 30 years. Another factor is the sharp rise in shrimp cultivation, which has created permanent saline water-logging in the region. Shrimp, which need sea water to grow, are a significant foreign-exchange earner and farmers have taken to building high mud walls around their farms to retain the saline sea water of the high tide. Over the past three decades, thousands of shrimp farms have sprung up in the region. Yet while sea water helps the shrimp farmers, it destroys all

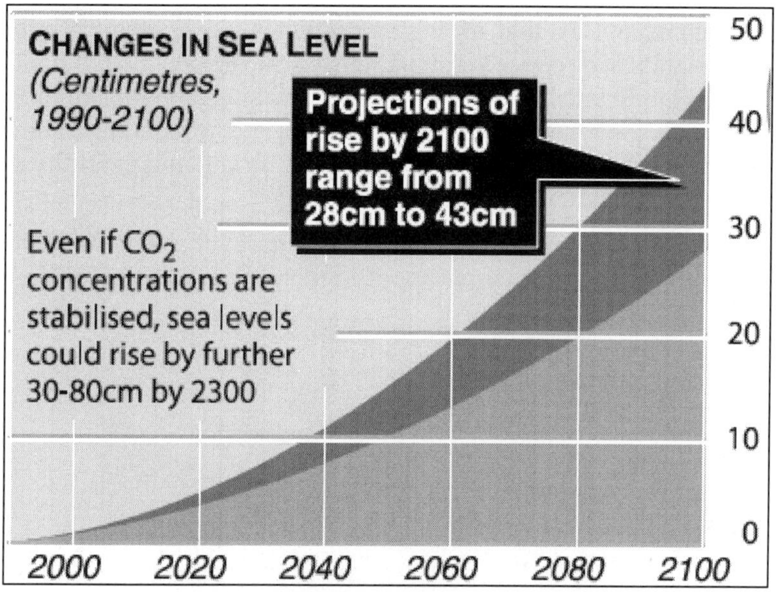

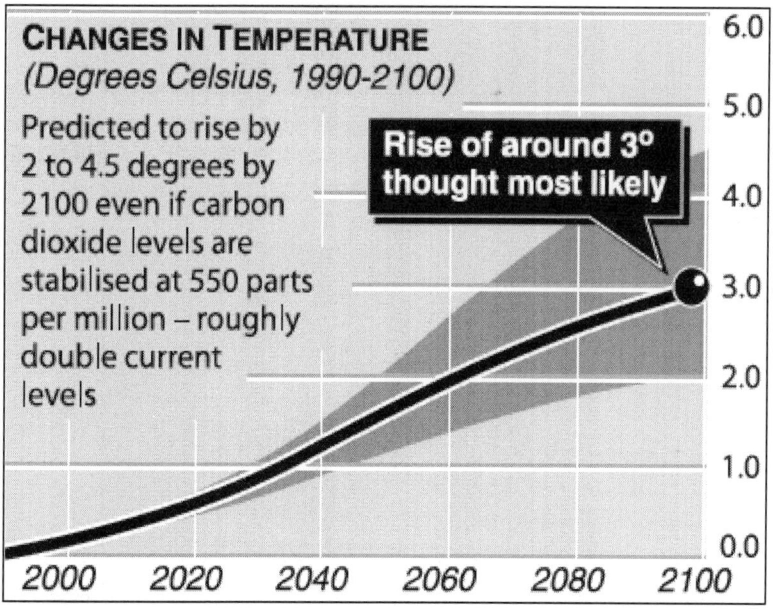

Figure 5.4: Changes in Sea Level (cm) and Temperature (°C)

other vegetation. "Farmers are more than compensated for their crop losses by growing shrimp in their erstwhile paddy fields. Shrimp fetches more money than rice," the government official said. But while accepting that an incursion of sea water into traditional croplands had reduced crop patterns in the region. "Only a handful of big landowners and the powerful are making money from shrimp. Poor people refuse to sell their land to shrimp farmers are victims of increasing salinity. Some people export shrimp and get filthy rich, but tens of thousands[17] (Plates 5.8–5.11).

A reporter Ane Ioran has been discussed an event of sea water incursion on www.climatefrontlines.org. Seawater incursion is a

Plate 5.8: Saline Sea Water Engulfs Rice Fields in the Southern District of Khulna, Bangladesh (*Source*: UNICEF Bangladesh)

Plate 5.9: People Use Bamboo Bridges to Cross Saline Puddles in Areas where Rising Sea Water Invades Villages (*Source*: UNICEF Bangladesh)

Plate 5.10: Shrimp Cultivators Hold Back Sea Water in Traditional Crop Fields (*Source*: UNICEF Bangladesh)

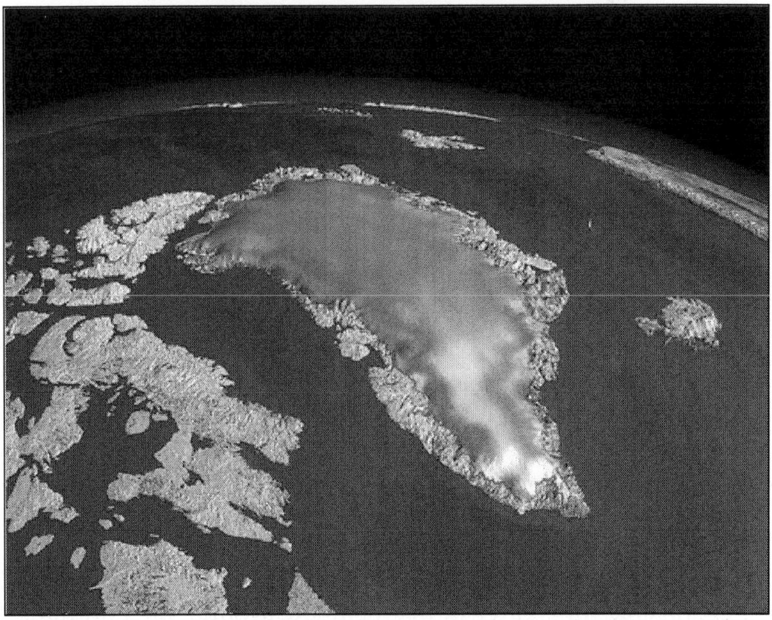

Plate 5.11: The IPCC 2007 Report Projected a Conservative Sea Level Rise of About 18–59 cm by the Year 2100

reality in Kiribati and it is a threat to the limited food trees and the major source of drinking water, the ground water. Before Kiribati submerges, seawater incursion will make atolls in the Republic of Kiribati unable to support life anymore and become uninhabitable.[18]

One of the more apparent consequences is the increased migration of salt water inland in coastal aquifers. Using two coastal aquifers, one in Egypt and the other in India, this study investigates the effect of likely climate change on sea water intrusion.[19]

5.1.2 Due to Man Kind Intervention

5.1.2.1 Increased Cropped Area/Intensity of Cultivation

Area under agriculture is declining; crop productivity is stagnated since last one decade, attributed to frequent occurrences of weather extremes. Intensive cultivation of crop is a necessity for food security of livelihood, which is possible through excessive use of inorganic fertilizer, insecticides pesticides, herbicides etc. These chemicals are responsible for degradation of soil health, changes of biodiversity, contamination of surface and ground water all these factors are involve in variation of micro climate.

5.1.2.2 Deforestation

Another cause of global warming is deforestation that is caused by cutting and burning of forests for the purpose of residence and industrialization. Deforestation is the conversion of forested areas to non-forest land for use such as arable land, pasture, urban use, logged area, or wasteland. Generally, the removal or destruction of significant areas of forest cover has resulted in a degraded environment with reduced biodiversity. In many countries, massive deforestation is ongoing and is shaping climate and geography. Deforestation results from removal of trees without sufficient reforestation, and results in declines in habitat and biodiversity, wood for fuel and industrial use, and quality of life (Plates 5.12 and 5.13).

5.1.2.3 Water-storage/Dams

Development of water storage structure/dames may also responsible for change of microclimate around surrounding area and also responsible for soil salinization and water logging problems in nearby areas.

5.1.2.4 Increasing Industrial Area

Industrial development is important for economic growth, employment generation and improvement in the quality of life. However, industrial activities without proper precautionary

Plate 5.12: Jungle Burned for Agriculture in Southern Mexico

Plate 5.13: Cutting of Trees Not Food for Climate

(*Source*: http://encyclopedia.thefreedictionary.com/deforestation)

measures for environmental protection are known to cause pollution and associated problems. If ecological and environmental criteria are forsaken, "industrialize and perish" will be the nature's retort. Now, there is a global consensus about the threat posed by the climate change. The disagreement is only, on how to go about altering human activities that unleash greenhouse gases, fuelling global warming. The recent report of the Intergovernmental Panel on Climate Change is the latest scientific assessment of the impact of global warming on human, animal and plant life. The culprit is greenhouse gases, notably carbon dioxide, methane and nitrous oxide. These are accumulating to unprecedented levels in the atmosphere as a result of profligate burning of fossil fuels, industrial processes, farming activities and changing land use. The greenhouse gases act like a blanket around the earth, trapping too much of the heat that would otherwise have escaped into space. The IPCC is a body of 2500 scientists that brings out reports, considered the last word on the Science of Climate Change. "Warming of the Climate System is unequivocal", says the IPCC in its latest report, pointing to the increased global, air and ocean temperatures, widespread melting of snow and ice and rising sea levels. If the introduction of these greenhouse gases continued to soar, global temperature could rise up by 2.4°C to 6.4°C by the end of the century, with far-reaching consequences for the climate, warned the IPCC. The report has given fresh impetus to finding solutions to the global warming problem.

5.2 Efforts to Overcome the Impact of Climate Change

To overcome the impact of climate change we should adopt land/water conservation agriculture methods suited to varied agro-climates. Increase crop diversity by inter-cropping and appropriate cropping systems/rotations/land use. Balanced use of biocides/chemicals. Increase forest area up to 33 per cent of the total geographical area. Increase carbon fixation in the soil by growing deep-rooted crops so as to decrease carbon foot–print. Use water judiciously (more crop/unit of water). Use less fossil fuel by using more solar/wind sources of energy. Educate farmers on the dangers of climate change. Disseminate meteorological/climate data/information on a large–scale. Suggest weather-based changes in cropping systems/land uses to sustain agricultural production. Encourage farmer groups to establish small weather observatories in their villages. Adopt use of soil-health cards widely for making fertilizer use decisions. Employ crop-weather models dynamically to advice farmers on improved animal/crop management for sustainable agriculture in a Decision Support System's framework.

India is a large developing country with nearly 700 million rural population directly depending on climate-sensitive sectors (agriculture, forests and fisheries) and natural resources (such as water, biodiversity, mangroves, coastal zones, grasslands) for their subsistence and livelihoods. Further, the adaptive capacity of dry land farmers, forest dwellers, fisher folk, and nomadic shepherds is very low[20]. Climate change is likely to impact all the natural ecosystems as well as socio-economic systems as shown by the National Communications Report of India to the UNFCCC[21].

5.2.1 What Can We Do?

§ Increase crop diversity by inter-cropping and appropriate cropping systems/rotations/land use: Area under agriculture is declining; crop productivity is stagnated since last one decade, attributed to frequent occurrences of weather extremes. If we have to adopt crop diversity, by using inter cropping, appropriate cropping systems, crop rotations and other scientific approaches for efficient land use to over-come the degradation of soil health, changes of biodiversity, contamination of surface and ground water by minimum utilization of inorganic fertilizers and chemicals (Plate 5.14).

- ☆ By adoption of scientific land/water conservation techniques suited to varied agro-climates we manage efficiently our natural resources and reduce the emission of methane, nitrous oxide etc. gases those are responsible for global warming.
- ☆ Balanced use of biocides/chemicals also reduce the degradation of soil health, changes of biodiversity, contamination of surface and ground water etc. enhance the safe nutritious food for secure livelihood.
- ☆ Increase forested area to 33 per cent of the total geographical area. Deforestation is responsible between 20–25 per cent of global greenhouse gas emissions. It is warned that a third of the State's biodiversity would vanish or would be close to extinction by 2030. There is urgent need for action plan in protecting natural forests at the State and National

Plate 5.14: Different Inter Cropping Systems to Mitigate the Impact of Climate Change for Security of Livelihood

Contd...

Plate 5.14–Contd...

levels to slow down global warming and to check extinction of forest species. In addition, tree planting should be on top priority under various schemes *viz.*, afforestation, agro-forestry, social forestry and farm level plantations as a part of carbon sequestration. Adoption of forest conservation, reforestation, afforestation and sustainable forest management practices can contribute to conservation of biodiversity, watershed protection, rural employment generation, increased incomes to forest dwellers and carbon sink enhancement.

☆ Increase carbon fixation in the soil by growing deep-rooted crops so as to decrease carbon foot–print: Deep rooted crops are helps in addition of more biomass in the soil and after decomposition of that root biomass may fix the significant quantity of carbon content in the soil and increase the productivity of soil in a eco-friendly system.

☆ Use water judiciously: more crop/unit of water: Technologies in rainwater harvesting, water conservation, judicious use of water and *in-situ* soil moisture conservation should be popularised to mitigate the effects of drought during summer when prolonged dry spell occurs if pre-monsoon showers fail.

☆ Use less fossil fuels as per need and contribute in reduction of carbon dioxide and carbon monooxide ultimately control the global warming. Efficient, fast and reliable public transport systems such as metro-railways can reduce urban congestion, local pollution and greenhouse gas emissions.

☆ Use more solar/wind sources of energy: Adoption of cost-effective energy-efficient technologies in electricity generation, transmission distribution, and end-use can reduce costs and local pollution in addition to reduction of greenhouse gas emissions.

☆ Educate farmers on the dangers of climate change: Adoption of participatory approach to forest management, rural energy, irrigation water management and rural development in general can promote sustained development activities and ensure long-term greenhouse gas emission reduction or carbon sink enhancement.

Plate 5.15: Stop Residue Burning to Save Carbon and Incorporate Crop Residue *in situ*

Plate 5.16: Awareness Programme of Farmers about Weather should be Conducted

Improved understanding of the exposure, sensitivity, adaptability and vulnerability of physical, ecological and social systems to climate change at regional and local level[22].

- Disseminate meteorological/climate data/information on a large–scale: Ministry of earth science is running a programme of integrated Agromet Advisory Service Bulletin under the supervision of IMD and NCMRWF through 130 Agromet Field Units (AMFUs) in all over India. All these AMFUs are prepared and disseminate district level bulletins on the basis of next 5 days weather forecast twice in a week every Tuesday and Friday. These bulletins are disseminated through Doordarsan, print and voice media, NGOs, ATMA, KVKs, Internet, ATIC etc.
- Suggest weather-based changes in cropping systems/land uses to sustain agricultural production: All the AMFUs are disseminated Integrated Agromet Advisory Service Bulletin based on weather forecasting at the district level involving multi-institutes and multi disciplinary scientists to mitigate the ill effects of climate change/variability on agriculture and life.
- Encourage farmer groups to establish small weather observatories in their villages: Recently IMD has started one day roving seminar on weather, climate and farmers and provision to distribute 25 plastic rain gauge to interested farmers of different villages along with sufficient training and awareness material at every AMFUs for encouragement and awareness of farmer groups about weather condition.
- Adopt use of soil-health cards widely for making fertilizer use decisions.
- Employ crop-weather models dynamically to advice farmers on improved animal/crop management for sustainable agriculture in a Decision Support System's framework.
- Stop residue burning and to save carbon and incorporate crop residues *in situ* (Plate 5.16).

5.3 References

1. Nettles, M. (2008) Step-wise changes in glacier flow speed coincide with calving and glacial earthquakes at Helheim Glacier, Greenland. Geophysical Research Letters, 35, doi:10.1029/2008GL036127.

2. Peterson, Thomas & Connolley, William & Fleck, John (2008) The Myth of the 1970s Global Cooling Scientific Consensus. American Meteorological Society. doi:10.1175/2008BAMS 2370.1.

   http://scienceblogs.com/stoat/Myth-1970-Global-Cooling-BAMS-2008.pdf.

3. "Science Past from the issue of (1959) Science News: p. 30. May 9, 2009. http://www.sciencenews.org/view/generic/id/43155/title/Science_Past_from_the_issue_of_May_9 per cent 2C_1959.

4. Erlich, Paul. (1968) "Paul Erhlich on climate change". Backseat driving.

   http://backseatdriving.blogspot.com/2005_07_01_backseatdriving_archive.html#112148592454360291.

5. Schneider S.H. (1972). "Atmospheric particles and climate: can we evaluate the impact of mans activities?". Quaternary Research 2 (3): 425–35. doi:10.1016/0033-5894(72)90068-3. [Precis Lay summary].

6. World's temperature likely to rise; The Times; 22 June 1976; pg 9; col A

7. Rasool, S.I.; Schneider, S.H. (1971) "Atmospheric Carbon Dioxide and Aerosols: Effects of Large Increases on Global Climate". Science 173 (3992): 138.

   doi:10.1126/science.173.3992.138

8. Keneth L. Denman and Guy Brasseur, *et al.* (2007) "Couplings between changes in Climate System and the Biogeochemistry, 7.5.3" (PDF). IPCC.

   http://www.ipcc.ch/pdf/assessment-report/ar4/wg1/ar4-wg1-chapter7.pdf.

Chapter 6

Global Climate Change Impact and Restoring Climate Order for Sustainable Food Security and Soil Health

Global warming is a reality on planet earth. If appropriate recommended measures are adopted in time, than we can reduce the overall impact of climate change and with the passes of time, we would be able to make earth cooler and harmonious environment for well being of human, animals and plant lives. Several tactics and practices are adopted in developed and developing countries to make the earth cool.

There should be two fold approaches to mitigate the climate stress–firstly by reducing greenhouse gas emission (GHGs), the main culprit of climate change and secondly, by adopting necessary farming practices like sustainable forestry systems, diversified cropping systems, carbon sequestration, recourse conserving technologies (RCTs) like zero tillage, minimum tillage or no till system, clean development mechanism (CDM), use biogas slurry, use or organic manure as a source of plant nutrients and introduction of resistant varieties to droughts, frost, insect pest and logging etc.

Soil management practices that usually improve soil organic matter include, (i) more complex crop rotation, especially those with high residue crops, (ii) intensive use of cover crops, (iii) use of variety of organic amendments, (iv) balanced fertilization, and reduced tillage. The various methods of carbon farming for capturing and holding carbon in soil includes, establishment of permanent vegetative cover as in the conservation reserve programs (pasture management), conservation tillage practices such as no-tillage; increased return of organic to soil through perennial crops and greater yields of annual crops and reduction of fallow periods, grazing management, biological farming, mulching and aforestation. Uses of improved farming practices helps in increasing the carbon pool of soils which have been lowered due to overexploitation and soil stress. Following the Kyoto protocol, many studies on carbon sequestration have been carried out or in process in African and Asian countries and found positive impact for cooling the earth. This research review discusses various adoptable and agricultural practices for earth cool.

Keywords: *Climate order, living planet, green house gases, conserving technologies.*

6.1 Global Climate Change Impact

Global green house (GHGs) emission due to human activities have gone since pre industrial times,with an increase of 70 per cent between 1970 to 2004. Carbondioxide is most important anthropogenic GHGs. Its annual emissions have grown between 1970 and 2004 by about 80 per cent. The largest growth of GHG emission between 1970 to 1994 has come from industry, energy supply, transport, forestry including deforestation, agricultural growth have been decreasing. Global GHGs have increased markedly as a result of human activities since 1750 and now far exceed pre-industrial values determined from ice cores spanning many thousands of years.

Figure 6.1 illustrates that the primary sources of the slow but steady increase in atmospheric carbon that is now occurring are fossil fuel combustion, which contributes approximately 5.5 gigatons (billion metric tons) of carbon per year, and land use changes, which account for another 1.1 gigatons. In contrast, the oceans absorb from the atmosphere approximately 2 more gigatons of carbon than they release and the earth's ecosystems appear to be accumulating another 1.2 gigatons annually. In all, the atmosphere is annually absorbing approximately 3.4 gigatons of carbon more than it is releasing.

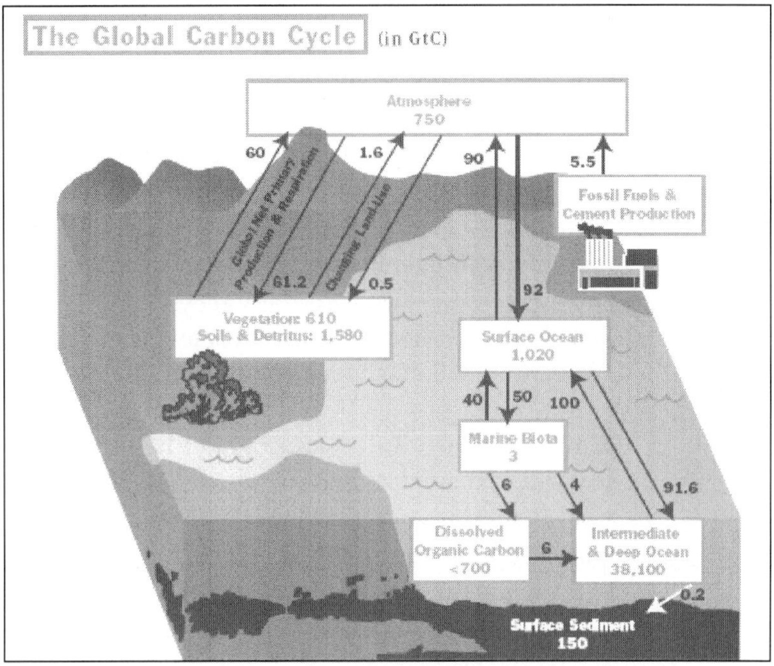

Figure 6.1: Global Carbon Cycle

(*Source*: www.metoffice.gov.uk/research/hadleycentre/models/ carbon_cycle/intro_global)

Global climate change is means an overall increase in temperature of universe. The eleven of the last twelve years (1995-2006) rank is warmest years in the instrumental records of global surface temperature (since 1850). The 100 year linear warming trend (1906-2006) of 0.74 °C is larger than the corresponding trend of 0.6 °C (1901 -2000). The linear warming trend over last 50 years from 1906 to 2005 (0.13 °C per decades). Observational evidence from all continents and most ocean shows that many natural systems are being affected by regional changes, particularly temperature increases. Climate change means variations in the climate in term of temperature, relative humidity, sunshine hours, wind velocity and other climatic parameters resulting changes in soil biodiversity, ground water level, soil degradation, erratic and uneven rainfall, frequent droughts and floods. Some common examples of climate change in Indian condition are:

- States like Bihar, Assam, and part of Karnataka are experiencing dry spell, whereas Southern Gujarat, Maharashtara, part of Bihar, Andhra Pradesh, Ladakh and Western Karnataka were hit by the floods.
- In 2007 alone, 17 million people had born the burnt of floods.
- During the year 2006, the Kashmir Valley is witness of most severe summer in three decades.
- Snowfall pattern of the Kashmir Valley Changes. During January and February no snowfall or less snowfall where as early snowfall in November and Late in March (2008-09).
- Charapuji known for highest rainfall had less in 2005. Mosinram experiences highest rainfall.
- Mumbai, for consequent 3-4 years, had heavy down pour, almost dipping the city.
- Unusual rainfall (60 cm. in 5 days, August19-23, 2006) in Barmer district of Rajasthan in 2006, was not recorded in the past 200 years.
- Evidences of loss of biodiversity (flora and fauna), genetic materials, soil microorganism at many places.

6.2 Impact of Climate Change on Agriculture and Systems

Climatic change is affecting all countries of the world either Asian, South East Asia or South Asia, European and African in big way. The poor country would be affected be in bigger way however they are contributing less in climate change. Himalayan ecosystem in which Kashmir Valley is situated is not a distant example from the impact of climate change. There is a direct link between the rise of global temperature (1 or 2°C) and damage to ecosystems. About 130 millian hectare land is under going different levels of degradation, namely water erosion (32.8 mha.), wind erosion (10.8 mha), salinisation (7.0 mha.), desertification (68.1 m.ha.), water logging (8.5 m.ha) and nutrient depletion (3.2 mha.). It has serious impact on the decreasing food productivity due to attack of insect and pest on crop, heavy rainfall, early or late maturity of crop. Small and marginal farmers with small land holding will be more vulnerability to climate change.

The resilience of many ecosystems is likely to be exceeded this century by an unprecedented combination of climate change and associated disturbances (flooding, drought, wildfire, insects, and ocean acidification) and other global change drivers (*e.g.* land use change, pollution, fragmentation of natural systems, over-exploitation of natural resources). Over the course of this century, net carbon uptake by terrestrial ecosystems is likely to peak before mid century and than weaken or reverse. Approximately 20-30 per cent of plant and animal species assessed so far are likely to be at increased risk of extinction if increase in global average temperature exceeds 1.5°C to 2.5°C. At lower altitude, especially in seasonally dry and tropical regions, crop productivity is projected to decrease for even small local temperature increases (1–2°C), which would increase the risk of hunger. The increase in atmospheric carbon concentration leads to further acidification of atmosphere and earth. Anthropogenic warming could lead to some extended impact depending upon magnitude of climate change. Uneven and erratic snowfall since last five years had disturbed Himalayan ecosystem.

Plate 6.1: Rise in Temperature would Shorten Life Cycle of Several crops

6.3 Climate Change and Water

It is supposed to suppress water resources. On regional scales, mountain snow, glaciers, and small ice caps play a crucial role in freshwater availability. Widespread mass losses from glaciers and reduction in snow cover over recent decades are projecting to accelerate throughout the 21 centuary, reducing water availability and hydropower potential, and changing seasonality of flows in regions supplied by melt water from major mountain range like Hindukush and Himalayas,where one-sixth of the world population currently lives. Changes in precipitation and temperature leads to changes in runoff and water availability. Runoff is projected to increase by 10-40 per cent by mid century at a higher latitudes and in some wet tropical areas, including populous areas in East and South East Asia and, decrease by 10-30 per cent over some dry tropics areas due to decrease in rainfall and higher rate of evapotranspiration. Drought affected area are projected to increase in extent, with the potential for adverse impacts on multiple sectors, *e.g.* agriculture, water supply, energy production and health. Large demand of water in urban areas for domestic purpose is on front and in rural areas for agriculture like irrigation of crops, rare ring livestock.

6.4 Tactics for Restoring Climate Order and Saving Living Planet–The Earth

There should be two fold approaches to mitigate the climate stress–firstly by reducing greenhouse gas emission (GHGs), the main culprit of climate change and secondly, by adopting necessary farming practices like sustainable forestry systems, diversified cropping systems, carbon sequestration, recourse conserving technologies (RCTs) like zero tillage, minimum tillage or no till system, clean development mechanism (CDM), use biogas slurry, use or organic manure as a source of plant nutrients and introduction of resistant varieties to droughts, frost, insect pest and logging etc. Uses of improved farming practices helps in increasing the carbon pool of soils which have been lowered due to overexploitation and soil stress.

6.5 Carbon Sequestration (CS)

Carbon sequestration refers to the storage of C into stable solid form. It occurs through direct and indirect fixation of atmospheric

CO_2. Direct soil C sequestration occurs by inorganic chemical reaction that converts CO_2 into soil inorganic carbon compounds such as Ca and Mg carbonates. Indirect plant C sequestration occurs as plants photosynthesise atmospheric CO_2 into plant biomass; subsequently some of the plant biomass is indirectly sequestered as soil organic carbon during decomposition process. The amount of carbon sequestered at a site reflects the long term balance between carbon uptake and release mechanism. Many best management practices have been proven to help in sequestering soil carbon are given below:

1. Restoration of degraded soils and ecosystems.
2. Adoption of recommended agricultural practices on prime land and
3. Retiring marginal agricultural lands to restorative land uses or converting to natural ecosystems.

With rapidly increasing population restoration of degraded soils and ecosystems is an important strategy. This strategy of restoration of degraded soils and ecosystems can enhance biomass production improves soil quality and increases the soil organic carbon (SOC) pool. Many soils of the tropics especially those in densely populated regions of Asia have lost a large proportion of their original SOC pool because of practices of mining soil fertility. There is a large potential of restoration of degraded soils in South East Asia which ranges from 18.3 to 35.0 Tera gram carbons per year (TgC/yr). These estimates are attainable potentials provided that regional governments adopt appropriate polices and implement plans to restore degraded soils through forestation, establishing planted fellows and improving grazing lands (Table 6.1). It is a major challenge that must be addressed in a coordinated and planned manner. In Haryana, India, it is reported a large increase in SOC content by reclamation of Sodic soils through growing *Prosopis juliflora* (Table 6.2). The SOC pool can also be increased by the addition of crop residue, maintenance of soil fertility through the integrated application of fertilizers, cattle manure and compost. In India use of NPK and FYM maintained the SOC at 15 gm/kg of soil for 25 year period compared with decline to 8.0 gm/kg with NPK alone and 5.0 gm/kg with no fertiliser use. Soil carbon sequestration in some soils in India from last 20 years is given in Table 6.3.

Table 6.1: Soil Carbon Sequestration through Restoration of Degraded Soils [17]

Degradation Process	*Area (m. ha)*	*Rate of Soil Organic Carbon Sequestration (Kg C/ha/yr)*	*Total Potential (Tera gram C/yr)*
Water erosion	81.2	100–150	8.1–12.2
Wind erosion	10.8	20.5	0.2–0.5
Desertification control			
1. Irrigated land	16.1	100–150	1.6–2.4
2. Rainfed	67	20–50	1.3–3.4
Salinization	33	200–500	6.6–16.5
Fertility depletion	10.9	100–150	1.1–1.6
Total			18.3–35.0

Table 6.2: Soil Organic Carbon Sequestration by Growing *Prosopis cineraria* in a Sodic Soil in Karnal [16]

Depth (cm)	*SOC Pool in Years After Planting (Mega gram C/ha)*			
	0	*5 yrs*	*7 yrs*	*30 yrs*
15	3.5	5	14.3	21.5
30	3.5	3.5	7.2	10.1
60	2.7	2.7	7.4	10.8
90	1.6	1.6	3.7	8.3
120	0.5	0.5	1.6	3.6
Total	11.8	13.3	34.2	54.3

6.6 Adoption of Recommended Agricultural Practices (RAPs)

Adoption of recommended Agricultural Practices (RAPs) can enhance the production of the above and below ground biomass and increase the SOC content. In comparison with other land uses *e.g.* permanent pastures, forests etc. Crop lands have lost most of their original SOC pool because of ploughing and susceptibility to erosion. Thus, there is a tremendous scope for improving SOC

Table 6.3: Soil Carbon Sequestration through INM for 20 Years in Some Soils of India [18]

Location	*Soil*	*Cropping System*	*SOC After 20 yrs.(gm/Kg)*			
			Initial	*Control*	*NPK*	*NPK+FYM*
Bhubneshwar	Inceptisol	Rice-rice	2.7	4.1	5.9	7.6
Pantnagar	Mollisol	Rice-wheat	14.8	5.0	9.5	15.1
Pantnagar	–	Rice-wheat-cowpea	14.8	6.0	9.0	14.4
Faizabad	Inceptisol	Rice-wheat	3.7	1.9	4.0	5.0
Barrakpore	–	Rice-wheat-jute	7.1	4.2	4.5	5.2
Palampur	Alfisol	Maize-wheat	7.9	6.2	8.3	12.0
Karnal	Alkali soil	Fallow-rice-wheat	2.3	3.0	3.2	3.5
Nagpur	Vertisol	Cotton-cotton	4.1	–	–	5.5
Trivandrum	Ultisol	Cassva	7.0	2.6	6.0	9.8

content through adoption of RAP's on crop lands. In addition to growing high yielding varieties within appropriate cropping sequence other components of RAPS are given below:

1. Using conservation tillage.
2. Soil fertility management.
3. Mulching/cover crops.
4. Enhancing rotational complexity.

If the above mentioned RMPs or components are followed than the South Asian countries have a potential of SOC sequestration equal to 11 to 22 Tera gram carbon per year (Tg C/yr)of which 8-16 Tg C/yr is in cropland of India (Table 6.4)

Table 6.4: Carbon Sequestration Potential through adoption of RAPs in South Asia [15]

Country	*Area (mha)*	*Rate of SOC Sequestration (Kg C/yr/ha)*	*Total Potential (Tera gram C)*
Afghanistan	7.9	50–100	0.1–0.2
Bangladesh	8.1	200–300	1.2–1.8
India	161.8	100–200	8.1–16.2
Iran	14.3	50–100	0.4–0.8
Nepal	3.1	300–500	0.7–1.2
Pakistan	21.5	50–100	0.5–1.0
Sri Lanka	0.9	300–500	0.2–0.4
Total	–	–	11.2–21.6

6.7 Land Uses Practices

Faulty land use practices like shifting cultivation, free-range grazing by cattle, growing crops along with the slope, cultivation of erosion permitting crops etc. may cause removal of top soil by erosion. Organic matter has low density than soil solids hence subjected to easily losses through wind and water erosion. It is clear that the OM loss under 3 per cent slope is around 46 kg/ha in Kerala. Cultivation of soil and consequent aeration stimulate more microbiological activities and promote the oxidation of organic matter *i.e.* increase the rate of disappearance of soil organic carbon. Intensive cultivation stimulates decomposition of Soil Organic Matter (SOM). Organic carbon status usually remains low in cultivated soils. It is clear that in all the soil zones, the organic matter content is very high in virgin soil.

The total biomass (t/ha) accumulated at the time of observation was maximum in natural forest *S. robusta* (NFSR), followed by plantation of *D. sissoo* (PDS), tea garden (TG), plantation of *T. arjuna* (PTA), mango-based agri-horticulture agroforestry system (AHAF), agricultural field (AF) and fallow land (FL) (Figure 6.2). However, while considering the rate of accumulation of biomass per unit area and per unit time with reference to the total biomass accumulated, land use systems based on annual crops exceeded the values compared to those having perennials as dominant components (Figure 6.3). Taking the amount of biomass accumulated per year, natural forest had an edge over the rest of land uses followed by PDS, TG, PTA and least in FL. Converting biomass into carbon stored into all the land uses followed by PDS, TG, PTA and least in FL. The Estimate of biomass accumulation under different land uses and their corresponding carbon stock are given in Figures 6.2 and 6.3.

6.8 Crop Rotations

Compared with monoculture cropping practices, multi crop rotations with two or three crops in a year can result in increased SOC contents (Table 6.5). This is because of addition of large amount of above ground as well as underground biomass in soil. Some crops grown in rotation leaves especially large quantities of residue, contributing greatly to the addition side of the gains-losses equation. Some crops, such as legumes, grasses, or grass legume forage crops, supply a lot of root biomass, which can contribute to residue. As compared to tomato inclusion of sunflower in pearl millet-potato cropping system enhanced SOC contents. This is further heightened due to introduction of leguminous green manuring crops like dhaincha (*Sesbania aculeate* L.) in pearlmillet-wheat-dhaincha, cowpea in sorghum cowpea, and clover in rice clover crop rotations.

6.9 Integrated Soil Fertility Management (ISFM)

Over the last few years, the concept of integrated soil management (ISM) and integrated plant nutrient management (IPNM) has been gaining acceptance. It advocates the careful management of nutrient stocks and flows in a way that leads to

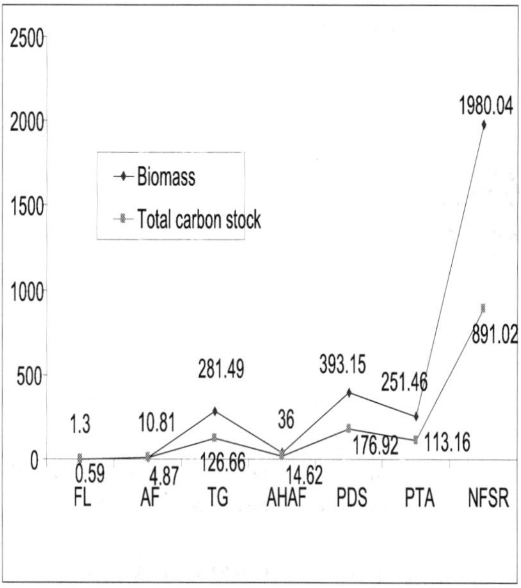

Figure 6.2: Biomass and Carbon Stock (t/ha) in Different Land Use

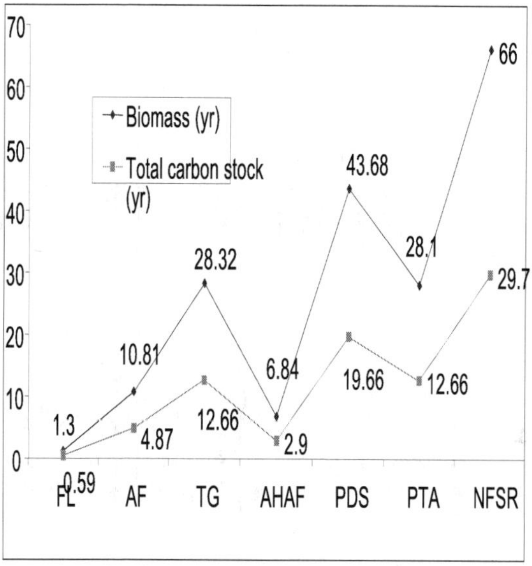

Figure 6.3: Biomass and Carbon Increment (t/ha/yr) Under Different Land Uses

Table 6.5: Effect of Crop Rotation on Soil Organic Carbon Contents in Soils of Sub-Humid, Semi- and Tropical Ecosystems

Cropping System	*Soil Organic Carbon ($g\ ka^{-1}$)*	*Microbial Biomass Carbon ($mg\ kg^{-1}$)*	*Total Nitrogen (per cent) Available Nitrogen ($kg\ ha^{-1}$)*	*References*
	Inceptisol, Hisar, Haryana, 6 years			
Pearmillet-Wheat-fallow	4.83	192	0.060#	1
Pearmillet-Fodder-cowpea-fallow	4.78	231	0.061	
Pearmillet-Potato-Tomato	5.04	221	0.063	
Pearmillet-Potato-Sunflower	4.54	184	0.059	
Pearmillet-Mustard-Sunflower	5.02	144	0.062	
LSD (p=0.05)	0.09	11	0.005	
	Vertisol, Hisar, Haryana, 2 years			
Sole Sorghum	6.21		230*	2
Sorghum-cowpea	7.07		233	
	Inceptisol, Hisar, Haryana, 5 years			
Rice-mustard	5.2	201.7	0.064#	3
Rice-wheat	6.0	270.0	0.055	
Rice-clover	7.2	241.8	0.083	
Sorghum-wheat	5.5	231.2	0.0061	
CD (p=0.05)	1.4	NS	0.001	

Total N in per cent, *: available N in $kg\ ha^{-1}$

CD: Critical difference; LSD: Least significant difference; NS: Not significant profitable and sustained production. ISM emphasises the management of nutrient flows, but does not ignore other important aspects of the soil complex, such as maintaining organic matter content, soil structure and soil biodiversity. Soil biodiversity reflects the mix and populations of diverse living organisms in the soil–the myriad of invisible microbes to the more familiar macro-fauna such as earthworms and termites. These organisms interact with one another and with plants and animals forming a web of biological activity. Environmental factors, including temperature, moisture, acidity and several chemical components of the soil affect soil biological activity. Clearly, for a productive sustainable agriculture, the complex interaction among these factors must be understood so that they can be managed as an integrated system.

Nitrogen availability can influence soil C levels in a variety of ways. It is clear that by increasing crop production, and thereby residue inputs, N fertilization can contribute to increased SOM contents. By increasing plant growth, fertilization can also lead to increased transpiration, drier soils and decreased decomposition rates. Results from many long term studies show a general tendency of increases in soil C with substantive additions of N, compared to zero or low N additions. Nitrogen additions can affect decomposition rates and C stabilization efficiency in other ways that contribute to higher SOM levels. Fog[4] reviewed 60 papers which reported zero or negative effects of N addition on decomposition rates. He offered several possible explanations for negative effects of N additions on decomposition, including the repression of lignolytic enzymes by ammonium and an increase in the amount of amino compounds which can act as precursors in the formation of recalcitrant humic compounds. At the microbial level, insufficient N can lead to lower yield efficiencies (*i.e.* more CO_2 respired per unit C assimilated[5]. Under such conditions addition of N could increase growth efficiency resulting in a higher proportion of C inputs retained in SOM.

In an analysis of long term plot studies with constant above ground C, with and without N fertilization, Paustain *et al.*[6] found that increased root residue inputs could not account for observed increases in soil C in the N fertilized treatments (Fig. 6.4). They also found that soil C:N ratio were lower in the unfertilized treatments, where straw or sowdust were added, suggesting that N limitation may have reduced C stabilization efficiency. In a study by Campbell

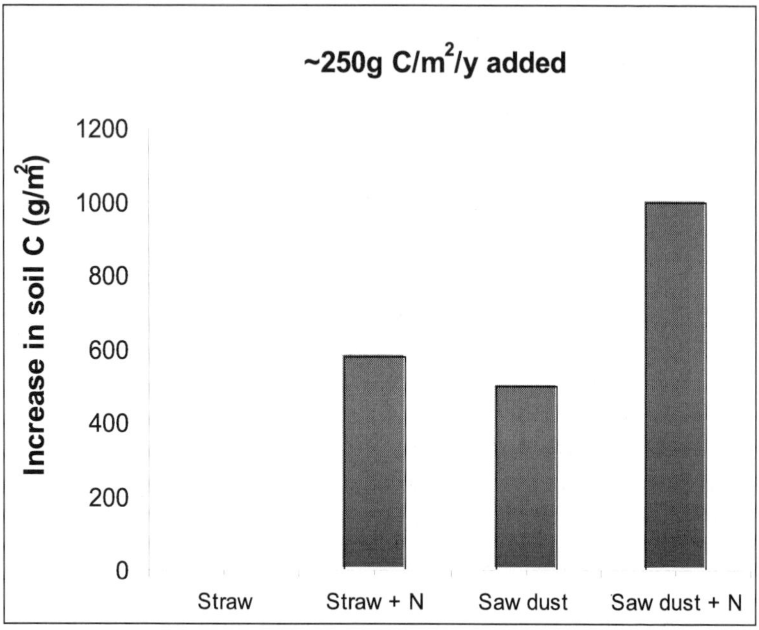

C:N ratio	9.1	9.6	9.9	10.2

Figure 6.4: N Addition Effect on the Net Change in Soil C Over the Duration of a 31 Years Experiment with Constant (~250 g cm^{-2} y^{-1}) of Straw or Sawdust Addition Aboveground Crop Residues were Removed from the Plots. N fertilized plots received rates equivalent to 80 kg N ha^{-1} y^{-1} as $Ca(NO_3)_2$

et al.[7], soil C was found to be similar in fertilized plots where straw was removed compared with fertilized plots where straw was retrained. This was despite the fact that C inputs were estimated to be greater in the treatment with straw retention.

In most field experiments it is difficult to partition the interacting and potentially conflicting, effects of N addition as soil C. However, when viewed in a broad sense, it is reasonable that, since C and N are the major constituents of SOM and their proportionality (*i.e.* C:N ratio) is relatively constant across a range of agricultural soils, then an adequate supply of N is required to build SOM. If inputs of these two elements are too much out of balance then the efficiency of soil C sequestration will be reduced. The balanced application of NPK (100 per cent or 150 per cent NPK) also showed higher accumulation of SOC over imbalanced use of fertilizers (100 per cent N & 100 per cent NP) in different cropping systems (maize-wheat-cowpea, rice-wheat-jute, maize-wheat, and soybean-wheat) over three decades under dissimilar climate and soil (Table 4.3)[8, 9, 10, 11]. Total organic carbon content in entire 0-45 cm soil profile in maize-wheat-cowpea cropping system followed the order: 150 per cent NPK + FYM > 150 per cent NPK > 100 per cent NPK > 100 per cent NP = 100 per cent N = 50 per cent NPK > control.

Chand[12] complied 200 applied integrated nutrients management (INM) practices for crops and cropping systems of various states of country, found that integration of nutrient sources both organic and inorganic results in sustainable yield and enriches SOC of soils without deterioration in soil texture and fertility. Similarly Chand[13] in a experiment on integrated nutrient management in mustard conducted at Udaipur, found that 50 per cent of recommended dose of NPK's coupled with FYM @ 10 t ha^{-1} + use of microbial inoculants (*Azotobactor* + PSB) helps in enriching yield of mustard besides improving soil physical, biological and chemical properties without negative effect on land. INM harmonizes soil, crop, water and other input in cropping lands by increasing use efficiencies (fertilizer, water or input) and improving post harvest SOC and nutrients.

6.10 Biological Management of Soil Fertility

It is central paradigm for the biological management of soil fertility is to utilise farmer's management practices to influence soil biological populations and processes in such a way as to achieve desirable effects on soil productivity[14]. Biological populations and processes influence soil fertility and structure in a variety of ways, each of which can have an ameliorating effect on the main soil-based constraints to productivity: symbionts such as rhizobia and mycrrhiza increase the efficiency of nutrient acquistition by plants; a wide range of fungi, bacteria, and animals participate in the process of decomposition, mineralization, and nutrient immobilisation and therefore influence the efficiency of nutrient cycles; soil organisms mediate both the synthesis and decomposition of soil organic matter and therefore influence cation exchange capacity, the soil N, S, and P reserve, soil acidity and toxicity; and soil water holding capacity; the burrowing and particle transport activities of soil fauna, and the aggregation of soil particles by fungi and bacteria, influence soil structure and soil water regime.

6.11 Role of Biodiversity

The role of soil biota/biodiversity in sustaining the productivity of agricultural systems a fundamental shift is taking place worldwide in agricultural research and food production in climate change scenario. In the past, the principal driving force was to increase the yield potential of food crops and to maximise productivity. Today, the drive for productivity is increasingly combined with a desire and even a demand for sustainability. Sustainable agriculture involves the successful management of agricultural resources to satisfy human needs while maintaining or enhancing environmental quality and conserving natural for future generations. Improvement in agricultural sustainability will require the optimal use and management of soil physical properties. Both rely on soil biological process and soil biodiversity. This implies management practices that enhance soil biological activity and thereby build up long-term soil productivity and health. Such practices are of major importance in marginal lands to avoid degradation, in degraded lands in need of restoration and in regions where high external input agriculture is not feasible.

6.12 Resources Conserving Technologies (RCTs)

RCTs are very important for increasing the carbon pool in soils. Zero tillage system offer minimum soil disturbances during sowing of crops. Zero tillage, raise bed planting, cover crops, crop residue management and mulching proves unique opportunities for restoration of soil organic carbon in agricultural lands[19,20,21]. Combating climate change in time, it is imperative to use renewable energy sources at domestic level. The uses of refrigerators shall be restricted to reduce the emission of gases.Carbon sequestration by composting, rising of green legumes and uses of manure is an important activity in agricultural production systems. Effort must be made at all levels of societies, institutions, NGOs and other youth forms. People must sensitize to use less combustive vehicles preference shall be given to CNG (Compressed natural gases) vehicles. The use of renewable energy sources like solar torch, solar batteries, solar water heating system must be recognised for energy saving and reducing GHG emission.

Plate 6.3: Awareness must be Created Among Farmers and Public Through on Spot Training and Two Way Conversations

A Healthy Human Life Depends on Healthy Soils and Quality Produces
For that I Appeal to Save Nature, Soil, Water, Energy and Resources
For Present as well as Future Generations of Mankind.

– Subhash Chand

6.13 Future Line of Work

- ☆ Soil carbon sequestration practices (SCSPs) must be documented area and region wise and popularised for maintenance of SOC in cultivable soils for soil quality enhancement and soil health restoration.
- ☆ There should be paradigm shift in land use practices (LUPs) from resource degrading to resource conserving technologies (RCTs) for possible areas of high intensive agriculture.
- ☆ There is a need for a commission for safer use of soils and for soil, water & plant analysis protocols.

☆ The awareness should be created about soil, plant and water resources through a series of *Kisan Melas*, farmer's camps, trainings, workshop, symposiums, conferences.

☆ Land use management practices for soil carbon sequestrations and farming carbon for mitigating global climate change must be harmonised and popularised for benefits of farmers besides improving soil fertility and land productivity.

☆ Fertilization, crop rotation and recommended practices shall be modified according to climate change impact to take more advantages for resource conservation (soil, water, and nutrient, input).

☆ A firm policy should be needed for climate change research and curricula's of schools, colleges must be updated to aware students in particulars and mass in general so they can play an important role in maintaining cool climate and harmonious environment.

☆ To make any program or policy successful in any nation or continent, it is imperative to implement in positive way, honesty and dedication.

☆ There is urgent need to create climate change sense in our daily life by utilizing resource efficiently to save energy in the form of input, electricity and fuel for future use.

☆ Climate change technology must be documented and spread for public use region wise, country wise to make earth neither coolest nor hottest. The earth planet environment should be maintained in normal for well being of mankind.

6.14 Key References and Resources

1. Duriasami, V.P., Perumal, R. and Mani, A.K. (2001). J. Indian Soc. Soil Sci. 49: 435-439.

2. Batra, L. and Rao, D.L.N. (2004). Under publication.

3. Andren, O. (1987). Decomposition in the field of shoots and roots of barley, lucern and meadow fescue. Swed. J. Agric. Res. 17: 113. Fog, K. The effect of added nitrogen on the rate of decomposition of organic matter. Biol. Rev. 1988. 63: 433.

4. Fog, K. (1998). The effect of added nitrogen on the rate of decomposition of organic matter. Biol. Rev. 63: 433.

5. Tempest, D.W. and Neijssd, O.M.(1984). The status of Y_{ATP} and maintenance energy as biologically interpretable phenomenon. Ann. Rev. Microbiol. 38: 459.

6. Paustain, K., Parton, W.J. and Persson, J. (1992). Medeling soil organic matter in organic amended and nitrogen fertilized long term plots. Soil Sci. Soc. Am. J. 58: 476.

7. Campbell, C.A., Lafond, G.P., Zentner, R.P. and Biederbeck, V.O. (1991). Influence of fertilizer and straw baling on soil organic matter in a thin black chernozem in western Canada. Soil Biol. Biochem. 23: 443.

8. Beri, V., Sidhu, B.S., Bahl, G.S. and Bhat, A.L. (1995). Soil Use Manage. 11: 51-54.

9. Manna, M.C., Swarup, A., Wanjari, R.H., Ravankar, H.N., Mishra, B., Saba, M.N., Singh, Y.V., Sahi, D.K. and Sarap, P.A. (2006). Field Crops Res. 93: 264-280.

10. Rudrappa, L., Purakayastha, T.J., Singh, D. and Bhandraray, S. Soil Till. Res. (2006). 88: 180-192.

11. Subehia, S.K., Verma, S. and Sharma, S.P. (2005). J. Indian Soc. Soil. Sci. 53: 308-314.

12. Subhash Chand. (2008). Integrated nutrient management for sustaining crop productivity and soil health. International Book Distributing Company, Lucknow, India, 112.

13. Subhash Chand. (2001). Integrated nutrient management in mustard, Ph.D Thesis, MPUAT, Udaipur, India.

14. Subhash Chand. and Pabbi, S. (2006). Vermicomposting in organic farming in souvenir of agriculture summit Organized by Ministry of Agriculture GOI and FICCI, Vigyan Bhavan, New Delhi, pp. 1-6.

15. Balesdent, J.C., Chenu and Balabane, M. (2000). Soil and Tillage Res. 53: 215-220.

16. Bhojvaid, P.P and Timmer, V.R. (1998). Forest Ecol. Management 106: 181-193.

17. IPCC. (2000). Land use, land use change, and forestry. A special report of 1PCC. Cambridge Univ. Press, U. K., 377.

18. Stan, D.W., Steve, A.S., Donald, L.R. and Charles, T.G. (2000). Environ. Sciences 30: 2091-2098.

19. Subhash Chand, S.K. Bansal, Tahir Ali, J. A Wani and M.R. Masih (2008) International symposium on Natural Resource Management for Sustainable Agriculture, ARS, Durgapura, Jaipur, pp. 1.4.51

20. Subhash Chand, A.R. Trag, Nasir A. Dar, Badrul Hasan and Geroge Ebert. (2010). Book of abstract, International Conference on Soil Fertility and Soil Productivity–Difference on Efficiency of Soils for Land Uses, Expenditure and Returns, pp. 228.

21. Subhash Chand and L.L. Somani. (2001). International conference on nature farming and ecological balance. CSSHAU, Hissar (Haryana). Abstracts. 1: 138.

22. Subhash Chand, Tahir Ali, Lal Singh and Uzma Bashir (2010). Global climate change and saving the living planate: The earth. Paper presented in International Conference at CBSH, GBPUAT, Pantnagar.

Chapter 7

Precision Farming for Natural Resource Management (NRM) and Sustainable Crop Production

It would be a simple matter to describe the earth's surface if it were the same everywhere. The environment, however, is not like that: there is almost endless variety [1].

– Webster and Oliver (1990)

Crop production involves a combination of practices revolving around soil, crop, climate and management factors. The factors affecting crop yields and environmental sensitivity vary in both space and time. Thus a new concept has arisen to manage the space-time continuum in crop production called precision agricultural management or precision agriculture or site specific management, which is concerned with the management of variability of agricultural resources in the dimensions of both space and time. Precision agriculture is the application of technologies and principles to manage special and temporal variability associated with all aspects of agricultural production for the purpose of improving crop performance and environmental quality. Success in precision agriculture is related to how well it can be applied to assess, manage, and evaluate the space time continuum in crop production. The agronomic feasibility of precision agriculture has been intuitive, depending largely on the application of traditional management recommendations at finer scales, although new approaches are appearing. Precision agriculture technique offers improved profits through increased yield as well as potential savings in input costs. Our analysis suggests prospects for current precision management increase as the degree of spatial dependence increases, but the degree of difficulty in achieving precision management increases with temporal variance. Thus, management parameters with high spatial dependence and low temporal variance will be more easily managed precisely than those with large temporal variance.

Keywords: *Precision farming, GIS, GPS, Remote sensing.*

Agriculture is the backbone of our country and economy, which accounts for almost 26 per cent of GDP and employs 70 per cent of the population. Though this is a rosy picture of our agriculture, how long will it meet the growing demands of the ever-increasing population. This is a difficult question to be answered, if we depend only on traditional farming. To meet the forthcoming demand and challenge we have to divert towards new technologies, for revolutionizing our agricultural productivity.

Green revolution succeeded in India to increase the farmer's income, yield of major crops and made India self-reliant in food production, with the introduction of high-yielding varieties and use of synthetic fertilizers and pesticides[1]. In the post-green revolution period agricultural production has become stagnant, and horizontal expansion of cultivable lands became limited due to burgeoning population and industrialization. In 1952, India had 0.33 ha of available land per capita, which is likely to be reduced to 0.15 ha by the end of year 2000. As the availability of land has decreased, application of fertilizers and pesticides became necessary to increase production. The major effect is that our agriculture became chemicalized. In this situation, it is essential to develop eco-friendly technologies for maintaining crop productivity.

Every component of production agriculture–from natural resources to plants, production inputs, farm machinery and farm operators are variable. The factors affecting crop yields and environment sensitivity vary in both space and time. Managing soil and crops in space and time is the sustainable management principle for the twenty-first century, a principle exemplified by farming by soils capes, managing zone within the field, and managing the non-crop period [2]. Thus a new concept has arisen to manage the space-time continuum in crop production called precision agricultural management or precision agriculture or site specific management, which is concerned with the management of variability of agricultural resources in the dimensions of both space and time. Natural, inherent variability between and within fields means that mechanized farming could traditionally only applying crop treatments for average soil, nutrient, moisture, weed and growth conditions. Necessarily, this has led to over and under applications of herbicides, pesticides, irrigation and fertilizers. Precision agriculture, explore the technological capabilities that enable it, assess its agronomic feasibility and environmental efficacy, and evaluate its performance to date relative to economic and social impacts. Precision agriculture technologies are being developed that can sense micro-site specific conditions in real time and can automatically adjust treatments to meet each site's unique needs.

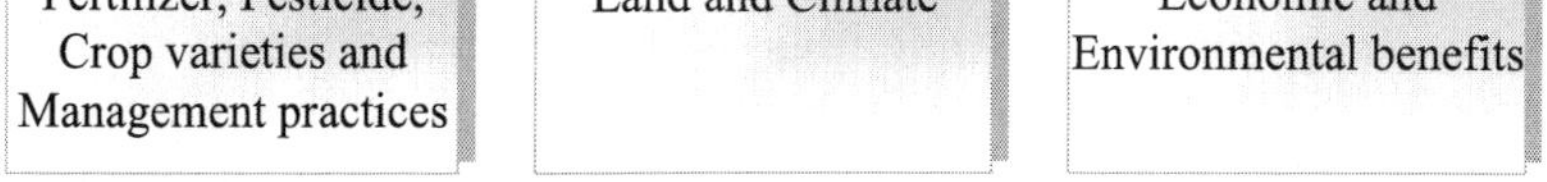

Resource to be managed

- Soil resource base
- Weather
- Plant genetics
- Crop diversity
- Machinery performance
- Physical, chemical, biological inputs, used in production of crop

Figure 7.1: Components of Precision Farming in Agriculture

7.1 Definition of Precision Agriculture

Currently, no precision agriculture system exist; rather, various components of traditional crop management systems have been addressed separately regarding their potential for site specific management, perhaps most notably soil fertility[3]. PF is a management philosophy or approach to the farm and is not a definable prescriptive system [4]. It is essentially more precise farm management made possible by modern technology. The variations occurring in crop or soil properties within a field are noted, mapped and then management actions are taken as a consequence of continued assessment of the spatial variability within that field. Development of geometrics technology in the later part of the 20th century has aided in the adoption of site-specific management systems using remote sensing (RS), GPS, and geographical information system (GIS). This approach is called PF or site-specific management[5][6]. It is a paradigm shift from conventional management practice of soil and crop in consequence with spatial variability. It is a refinement of good whole field management, where management decisions are adjusted to suit variations in resource conditions. PF requires special tools and resources to recognize the inherent spatial variability associated with soil characteristics, crop growth and to prescribe the most appropriate management strategy on a site-specific basis. It offers a potential step change in productive efficiency[4]. Fundamentally, PF acknowledges the conditions for agricultural production as determined by soil, weather and prior management across space and over time [7]. Considering this inherent variability, management decisions should be specific to time and place, rather than rigidly scheduled and uniform.

Conventional agriculture is practiced for uniform application of fertilizer, herbicide, insecticides, fungicides and irrigation, without considering spatial variability. To alleviate the ill-effects of over and under usage of inputs, the new concept of PF has emerged. Site-specific management to spatial variability of farm is developed to maximize crop production and to minimize environmental pollution and degradation, leading to sustainable development. The recommendations of production inputs for each variable portion of the field could be adjusted to optimize output according to the agronomic, economic[5] and environmental goals through minimization of production cost.

Today, farmers are adopting individual components of precision agriculture on the farm but a distinctive precision farming system has not yet evolved. Technological developments continue to occur and as a result of ongoing research a better understanding of underlying processes is being developed but a true system has not emerged. Therefore, any definition of precision agriculture can at best be considered only operational. Since the mid-1980s, a host of terms have been used to described to concept of precision agriculture, including farming by the foot[8], farming by soil[6][9], variable rate technology (VRT) [10], spatially variable, precision, prescription, or site-specific crop production [11], and site-specific management[12]. A recent report of a National Research Council, Board on Agriculture Committee defined precision agriculture as "a management strategy that uses information technologies to bring data from multiple sources to bear on decisions associated with crop production" (NRC, 1997; p. 17). On the basis of above information a complete definition of precision agriculture is as: "Precision agriculture is the application of technologies and principles to manage spatial and temporal variability associated with all aspects of agricultural production for the purpose of improving crop performance and environmental quality".

In the view of another scientist precision farming is "art and science of utility advanced technologies for enhancing crop yield while minimizing environmental threat to the planet"[13].

7.2 Concept of Precision Farming

The two basic concepts in precision farming are

7.2.1 Assessing Variability

In precision farming, inputs are to be applied precisely in accordance with the existing variability. Therefore, assessing the in-field variability is very crucial and first step of precision agriculture. Spatial variability of all the determinants of crop yield should be well recognized, adequately quantified and properly located. Construction of condition maps on the basis of variability is a critical component of precision farming. Condition maps can be generated through survey, point sampling and interpolation, remote sensing (high resolution), and modeling.

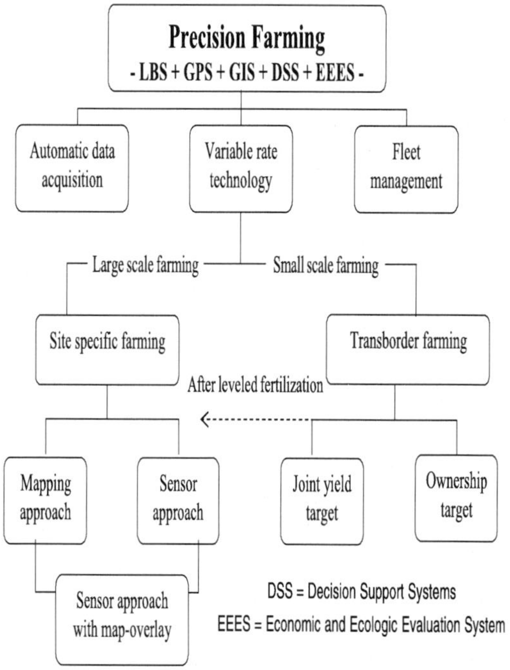

Figure 7.2: Systematic Integration of Precision Farming into Existing Agricultural Structures

7.2.2 Managing Variability

After assessing the spatial variability, the same can be taken care through precision land leveling, variable rate technology, site-specific planting, site-specific nutrient management, precision water management, site-specific weed management and precision pest and disease management etc. These precision farming practices aims at managing the variability by applying or making farm inputs available only in required quantities at particular time and at specific locations.

7.3 Technology Used for Precision Farming

The new tools applicable to this Precision Farming are the advances in electronics and computers such as RS, GPS and GIS. Technologies used in PF cover three aspects such as data collection, analysis or processing of recorded information and recommendations based on available information.

7.3.1 Mapping

The generation of maps for crop and soil properties is the most important and first step in PF. These maps will measure spatial variability and provide the basis for controlling spatial variability. Data collection occurs both before and during crop production and is enhanced by collecting precise location coordinates using the GPS. The data collection technologies are grid soil sampling, yield monitoring, RS and crop scouting [7]. During crop production, the data are collected through sensing instruments such as soil probes, electrical conductivity and soil nutrient status. Optical scanners are used to detect soil organic matter and to recognize weeds [7]. Then these data generated through mapping are recorded and stored in a computer system for future action and generated maps used for acquisition of information and for making strategic decisions to control variability. Mapping can be done by RS, GIS and manually during field operations.

7.3.2 Manual Mapping during Field Operations

Measurements may also be taken during field operations by the farmers. The most common measurements during field operation are yield recording and soil properties during tillage. Manual measurement has also been done for soil sample, pest infestation and other crop problems [14]. These measurements are performed at a specific time and usually provide the most accurate and useful information. Accurate and reliable sensors are needed for the conversion of physical and biological quantities into electronic value. Mapping also requires an accurate locator to establish the geographical location of the quantities measured, for which differential GPS (DGPS) is very useful in PF.

7.3.3 Remote Sensing (RS)

It is the acquisition of information about an object from a distance, with precision, without coming into contact with the same. Although the use of RS is a decade old, its relevance to agriculture in spatial variability management is relatively new. RS measures visible and invisible properties of a field or a group of fields and converts point measurements into spatial information, to monitor temporally dynamic plant and soil conditions. The visual observations are recorded through a digital notepad and geo-referenced to GIS database, the most commonly used RS device, but aerial photography and video-graphy are also used in PF [5]. Satellite RS has provided a tool for acreage estimation one month in advance, with more than 95 per cent accuracy and in mono-crop area yield estimation with more than 90 per cent accuracy ten days in advance [15]. These images allow mapping of crop, pest and soil properties for monitoring seasonally variable crop production [16], stress, weed infestation and extent [5] within a field.

RS can be used for PF in a number of ways for providing input supplies and variability management through decision support system. The point data of soil test results can be translated into spatial coverage based on geo-statistical interpolation [5], which gives chemical properties of the soil, nutrient status, organic matter, salinity, moisture content, etc. This information on spatial variability can be used with other geo-references to identify both seasonally stable and variable units, based on which management strategies can be developed. Space technology combined with satellite RS and communication provides valuable, accurate and timely information like early warning, occurrence, progressive dangers, damage assessment, quick dissemination of information regarding disaster and decision support to mitigate it [15]. Recent developments in remote sensing technology have led to significant improvements of the spatial and spectral resolution of commercially available aircraft or satellite–based remote sensing imagery.

7.3.4 Geographic Information System (GIS)

A Geographical Information System has the capability to capture, store, manipulate, analyze and display spatial information and related attributes. The graphics data base in the GIS contains all of the locational information relating to mapped features. The attribute data base contains all of the descriptive information relating to the mapped features. Use of GIS in agriculture has increased because of misuse of resources like land, water, etc. GIS is the principal technology used to integrate spatial data coming from various sources in a computer [7]. GIS techniques deal with the management of spatial information of soil properties, cropping systems, pest infestations and weather conditions. This is primarily an intermediate step because it combines the data collected at different times based on sampling regimes, to develop the subsequent decision technologies such as process models, expert systems, etc.

In the new millennium, GIS-aided techniques are indeed needed for sustainable food production and resource utilization, without further depletion of the environment. GIS technology will help the farmers and scientists in decision making, as precise information on field will be readily available. GIS techniques make weed control, pest control and fertilizer application site-specific, precise and effective; it would also be very useful for drought monitoring, yield

Application of GIS and modeling in agronomy and natural resources management research

- ☆ Atmospheric modeling
- ☆ Climate change, sensitivity and/or variability studies
- ☆ Agro-ecological characterization and zonation
- ☆ Regional risk analysis
- ☆ Scenario modeling and impact assessment
- ☆ Hydrology, water quality, water pollution
- ☆ Spatial yield calculation-regional, global
- ☆ Precision farming

Figure 7.3: Area for Application of GIS and Modeling in Agronomy

estimation, pest infestation monitoring and forecasting[15][17][18]. GIS coupled with GPS, microcomputers, RS and sensors is used for soil mapping, crop stress, yield mapping, estimation of soil organic matter [18] and available nutrients. In combination these technologies have brought out rapid changes in data collection, storing, processing, analysis and developing models for input parameters[17].

7.3.5 Global Positioning System (GPS)

The GPS technology enables precision agriculture because all phases of precision agriculture require positioning information. GPS is able to provide the positioning in a practical and efficient manner[19]. GPS was developed by the US military and later permitted for restricted civilian use. Expensive, high-precision differential GPS (DGPS) systems are available that achieve centimeter accuracies[20], allow for automated machinery guidance[21][19] and kinematic mapping of topography[22], and are useful in the creation of digital elevation models needed for terrain analysis[23] [24]. While the GPS signal is ubiquitous, there have been problems in making available GPS at the needed precision for agriculture[25]. Additionally, some GPS receivers are susceptible to unwanted interfering signals from a variety of sources, including farm machinery, making the receiver useless in navigation or positioning. Some interferences can be overcome in the design of the GPS receiver. The GPS can be used in two modes; single receiver mode and differential mode (DGPS) using two receivers. Single receiver collects the timing information and processes it into position. This system is the cheapest and easiest, but its accuracy suffers due to introduced positional errors. In the differential mode (DGPS), one receiver is mounted in a stationary position; usually at farm office while the other is on the machine/ implement [26].

Regardless of problems, DGPS has greatly enabled precision agriculture. Of great importance for precision agriculture, particularly for guidance and for digital elevation modeling, position accuracies at the centimeter level are possible in DGPS receivers that use carrier phase in combination with DGPS[20][19]. Accurate guidance and navigation systems will allow for farming operations not currently in use, including field operations at night when wind speeds are low and more suitable for spraying and the use of night tillage to reduce the light induced germination of certain weeds[27].

7.3.6 Modeling

Modeling is proposed as an important tool in precision agriculture to simulate spatial and temporal variation in soil properties [28], pests [29], crop yield [30] and environmental performance of cropping systems [31]. Models have been developed and calibrated for specific purposes but have not been used extensively in spatial prediction. A major problem of models is the availability of inputs needed to run them. A major advantage of models is their ability, once calibrated to simulate the temporal component of crop production. This capability should allow models designed to account for spatial variability to evaluate different precision management scenarios that would otherwise be prohibited by time and cost considerations. The application of models to the simulation of the space-time continuum of crop production is a critical research need [32].

7.4 Commercialization of PF

The interest in PF and its introduction has resulted in a gap between the technological capabilities and scientific understanding of the relationship between the input supplies and output products. Development of PF has been largely market-driven, but its future growth needs collaboration between private and public sectors. The private sector has to take up the responsibility of market development, product credibility and customer satisfaction. Whereas, the public sector needs to coordinate the activities involved in developing and implementing PF, by providing support programmes to achieve the objectives [5]. Linkages between government, university and corporate sectors are essential to facilitate the transfer and acceptance of technology by end-users. The potential of this technology has already been demonstrated, but in practice, meaningful delivery is difficult as it needs large-scale commercial application to realize the benefits.

7.5 Indian Perspective

The green revolution has not only increased productivity, but it has also several negative ecological consequences such as depletion of lands, decline in soil fertility, soil salinization, soil erosion, deterioration of environment, health hazards, poor sustainability of agricultural lands and degradation of biodiversity. Indiscriminate use of pesticides, irrigation and imbalanced fertilization has threatened sustainability[33] [15]. On the other hand, issues like declining use efficiency of inputs and dwindling output–input ratio have rendered crop production less remunerative. According to CGIAR, 'Sustainable agriculture is the successful management of resources to satisfy the changing human needs, while maintaining or enhancing the quality of environmental and conserving natural resources'.

The new technology should be highly productive, cost-effective and ecologically sustainable. In the present context, maintenance of ecological balances through precise and site-specific management is most desirable. The concept of PF may be appropriate to solve these problems, though it looks unsuitable to Indian conditions; but it is not impossible to adopt. Research efforts are needed to find out its applicability in the Indian agricultural scenario.

7.6 Site Specific Nutrient Management (SSNM) in Intensive Rice Systems in Asia

7.6.1 Major Components

- Modeling of crop nutrient requirements based on physiological requirements and an economically efficient yield target.
- Field-specific estimation of the indigenous supply of N, P, & K.
- Modeling of recovery efficiencies of N, P and K.
- Estimation of P & K balance to sustain soil P & K reserves without depletion.
- Dynamic adjustment of N application based on monitoring of the plant N status during critical periods of the growth *i.e.* using a chlorophyll meter or a leaf colour chart.

7.6.2 Conditions where Precision N Management will be Profitable or Beneficial to the Environment

- Where N inputs are high.
- Where residual N is temporarily stable and/or high residual N is predictable from the yield of the previous crop, *e.g.* low yields in the previous year.
- Where crop quality is affected by excess N in soil.
- Where crop yield spatial variability is high and predictable.

- ☆ Where net mineralization is high and consistently related to soil and landcape properties.
- ☆ Where N application is not restricted in time.
- ☆ Where leaching potential is high and time spatially variable to or during the crop N uptake period of plant growth.
- ☆ Where variation in topographic position regulates N availability or yield.

7.6.3 Conditions not Favouring Precision Nitrogen Management

- ☆ Where the previous crop exceeds yield goal.
- ☆ Where large-scale leaching events occur prior to the growing season.
- ☆ Where precision N management has successfully reduced spatial variation in yield and residual N.
- ☆ Where there is a strong temporal component of spatial variability.

7.7 References

1. Webster, R. and Oliver, M.A. (1990) "Statistical Methods in Soil and Land Resource Survey." Oxford Univ. Press, New York.
2. Pierce, F.J. and Lal, R. (1991) Soil management in the 21st century. In "Soil management for sustainability" (R. Lal and F.J. Pierce, Eds.), pp 175-180 Soil and Water Conservation Society, Ankeny, IA.
3. Lowenberg-DeBoer, J. and Swinton, S.M. (1997) Economics of site-specific management in agronomic crops. In "The State of Site-Specific Management for Agriculture" (F.J. Pierce and E.J. Sadler, Eds.), ASA Miscellaneous Publication. pp. 369-396. ASA, CSSA and SSSA, Madison, WI.
4. Dawson, C.J. (1997) *Precision agriculture* (ed. Stafford, J.V.). BIOS Scientific Publishers Ltd. Vol. 1. pp. 45-58.
5. Brisco, B.; Brown, R.J.; Hirose, T.; McNairn, H. and Staenz, K. (1998) *Can. J. Remote Sensing*. 24. 315-327.

6. Carr, M.P.; Carlson, G.R.; Jacobsen, J.S.; Nielsen, G.A. and Skogley, E.O. (1991) Farming soils, not fields: A strategy for increasing fertilizer profitability. *J. Prod. Agric.* 4, 57-61.

7. Heimlich, Ralph. April (1998) *Agric. Outlook.* 19-23.

8. Reichenberger, L. and Russnogle, J. (1998) Farm by the foot. Farm J. 113, 11-15.

9. Larson, W.E. and Robert, P.C. (1991) Farming by soil. *In* "Soil Management for Sustainability" (R. Lal and F.J. Pierce, Eds.), pp.103-112. Soil and Water Conservation Society, Ankeny, IA.

10. Sawyer, J.E. (1994) Concepts of variable rate technology with considerations for fertilizer application. *J. Prod. Agric.* 7, 195-201.

11. Schueller, J.K. (1991) In-field site-specific crop production. *In* "Automated Agriculture for the 21st Century. Proceedings of the 1991 Symposium, Chicago, IL, 16-17 December, "ASAE Publ. No. 11-91, pp. 291-292. ASAE, St. Josefph, MI.

12. Pierce, F.J. and Sadler, E.J. (Eds.). (1997) "The State of Site-Specific Management for Agriculture." ASA Miscellaneous Publication. ASA, CSSA and SSSA, Madison, WI.

13. Gill, M.S. (2006).

14. Stafford, J.V., Lebars, J.M. and Ambler, B. (1996) Comput. Electron. Agric. 14. 234-247.

15. Singh, K.K. and Shekhawat, M.S. (2000) Farmer and Parliament. 10-13.

16. Moran, M.S.; Vidal, A.; Troufleau, D.; Inoue, Y. and Mitchell, T. (1997) Remote Sensing Environ. 61. 96-109.

17. Reddy, G.P. and Anand, P.S.B. June (2000) Yojana. pp. 35-36.

18. Biswas, C. and Subba Rao, A.V.M. June (2000) *Yojana.* 24-25.

19. Tyler, D.A.; Roberts, D.W. and Nielsen, G.A. (1997) Location and guidance for site-specific management, In "The State of Site-Specific Management for Agriculture" (F.J. Pierce and E.J. Sadler, Eds), ASA Miscellaneous Publication pp. 161-181, ASA, CSSA and SSSA, Madison, WI.

20. Lange, A.F. (1996) Centimeter accuracy differential GPS foe precision agriculture applications. *In* "Proceedings of the Third

International Conference on Precision Agriculture, Minneapolis, MN, 23-26 June 1996" (P.C. Robert, R.H. Rust and W.E. Larson, Eds.), ASA Miscellaneous Publication. pp. 675-680. ASA, CSSA and SSSA, Madison, WI.

21. O'Conner, M.; Bell, T.; Elkaim, G. and Parkinson, B. (1996) Automatic steering of form vehicles using GPS. *In* "Proceedings of the Third International Conference on Precision Agriculture, Minneapolis, MN, 23-26 June 1996" (P.C. Robert, R.H. Rust and W.E. Larson, Eds.), ASA Miscellaneous Publication. pp. 767-777. ASA, CSSA and SSSA, Madison, WI.

22. Clark, R.L. (1996) A comparison of rapid GPS techniques for topographic mapping. *In* "Proceedings of the Third International Conference on Precision Agriculture, Minneapolis, MN, 23-26 June 1996" (P.C. Robert, R.H. Rust and W.E. Larson, Eds.), ASA Miscellaneous Publication. pp. 651-662. ASA, CSSA and SSSA, Madison, WI.

23. Bell, J.C.; Butler, C.A. and Thompson, J.A. (1995) Soil-terrain modeling for site-specific management. In " Proceedings of the Second International Conference on Site-Specific Management for Agricultural Systems, Bloomington/Minneapolis, MN, 27-30 March 1994" (P.C. Robert, R.H. Rust and W.E. Larson, Eds.), ASA Miscellaneous Publication, pp. 209-227. ASA, CSSA and SSSA, Madison, WI.

24. Moore, I.D.; Gessler, P.E.; Neilsen, G.A. and Peterson, G.A. (1993) Soil attribute prediction using terrain analysis. Soil Sci. Soc. Am. J. 57, 443-452.

25. Saunders, W.P., Larscheid, G.; Blackmore, B.S. and Stafford, J.V. (1996) A method for direct comparison of differential global positioning systems suitable for precision farming. In "Proceedings of the Third International Conference on Precision Agriculture, Minneapolis, MN, 23-26 June 1996" (P.C. Robert, R.H. Rust and W.E. Larson, Eds.), ASA Miscellaneous Publication. pp. 663-647. ASA, CSSA and SSSA, Madison, WI.

26. Yadav, R.L. and Rao, Subba A.V.M. (2000) In: National Symposium "Agronomy: Challenges and Strategies for the New Millennium", November 15-18, 2000, GAU, Junagadh. Indian Society of Agronomy. pp 3-4.

27. Hartmann, K.M. and Nezadal, W. (1990) Photocontrol of weeds without herbicides. *Naturwissenschaften* 77, 158-163.

28. Verhagen, J. and Bouma, J. (1997) Modeling soil variability. In "The State of Site-Specific Management for Agriculture" (F.J. Pierce and E.J. Sadler, Eds), ASA Miscellaneous Publication pp. 55-67, ASA, CSSA and SSSA, Madison, WI.

29. Kropff, M.J. and Lotz, L.A.P. (1997) Modeling for precision weed management. *In* "Precision Agriculture: Spatial and Temporal Variability of Environmental Quality" (J.V. Lake, G.R. Bock and J.A. Good Eds.), pp. 182-200. Wiley, New York.

30. Barnett, V.; Landau, S.; Colls, J.J.; Craigon, J.; Mitchell, R.A.C. and Payne, R.W. (1997) Predicting wheat yields: The search for vailid and precise models. In "Precision Agriculture: Spatial and Temporal Variability of Environmental Quality" (J.V. Lake, G.R. Bock and J.A. Goode. Eds.). pp. 79-92. Wiley, New York.

31. Verhagen, A.; Booltink, H.W.G. and Bouma, J. (1995) Site-specific management: Balancing production and environmental requirements at farm level. *Agric. Systems* 49, 369-384.

32. Sadler, E.J. and Russell, G. (1997) Modeling crop yield for site-specific management. *In*. "The State of Site-Specific Management for Agriculture" (F.J. Pierce and E.J. Sadler, Eds.), ASA Miscellaneous Publication. pp. 69-79. ASA, CSSA and SSSA, Madison, WI.

33. Ghosh, S.k.; Murthy, K.M.D.; Ramesh, G. and Palaniappan, S.P. (1999) *Employment news*. 222. 1-2.

34. Subhash Chand and Lal Singh (2010) Precision farming: A tool for future agriculture. Agriculture Extension Review. Ministry of Agriculture, New Delhi.

Chapter 8

Resource Conservation Technology for Sustainable Food Security

Depleting soil organic carbon status, decreasing soil fertility and reduced factor productivity are major issues for resource conservation technologies to sustain food production of the country. These evidences indicate that rice-wheat system has weakened the natural resource base. The indiscriminate use, rather misuse, of natural resources especially water has led to the groundwater pollution as well as depletion of ground water resources. If we continue to exploit the natural resources, the productivity and sustainability is bound to suffer. Therefore, in order to meet the aim of sustainable yields over time it is the need of the hour to avoid further degradation of the natural resources. Moreover, in the face of World Trade Organization (WTO) regime, we must produce at lower cost to be competitive in the international market being India already a surplus state in food-grain production. To meet these needs, the agricultural system must develop cost effective technologies suitable for harnessing the untapped potential especially in the north eastern parts of the Indo-gigantic Plains (IGP). The resource conservation technologies to economies on cost of production and need for conservation agriculture has assumed lot of significance. The article discusses the various resources conserving technologies for wheat, rice and for rice-wheat cropping system. The possibilities of RCTs to reclaim alkali soils have been also discuss. Various new machines have been described for better resource conservation in crop cultural practices.

Keywords: *Resource conservation technology, Conservation agriculture, Zero tillage, Rice-wheat cropping system, Laser land leveling.*

8.1 Conservation Agriculture (CA)

Conservation tillage methods are much more than just reducing the mechanical tillage. In a soil that is not tilled for many years, the crop residues remain on the soil surface and produce a layer of mulch. This layer protects the soil from the physical impact of rain and wind but it also stabilizes the soil moisture and temperature in the surface layers. Thus this zone becomes a habitat for a number of organisms, from larger insects down to soil borne fungi and bacteria. Those organisms generate the mulch, incorporate and mix it with the soil and decompose it so that it becomes humus and contributes to the physical stabilization of the soil structure. At the same time this soil organic matter provides a buffer function for water and nutrients. Larger components of the soil fauna, such as earthworms, provide a soil structuring effect producing very stable soil aggregates as well as uninterrupted macropores leading from the soil surface straight to the subsoil and allowing fast water infiltration in case of heavy rain events. This process carried out by the edaphon, the living component of a soil, can be called "biological tillage". However, biological tillage is not compatible with mechanical tillage and with increased mechanical tillage the biological soil structuring processes will disappear. Certain operations such as mould board or disc ploughing have a stronger impact on soil life than others as for example chisel ploughs. Most tillage operations are, however, targeted at a loosening of the soil which inevitably increases the oxygen content in the soil leading to mineralization and thus to a reduction of the soil organic matter which is, at the same time substrate for soil life. Thus agriculture with reduced mechanical tillage is only possible when soil organisms are taking over the task of tilling the soil. This, however, leads to other implications regarding the use of chemical farm inputs. Synthetic pesticides and mineral fertilizer have to be used in a way that does not harm soil life.

As the main objective of agriculture is the production of crops, changes in the pest and weed management become necessary. Burning of plant residues and ploughing of the soil is mainly considered necessary for phyto-sanitary reasons controlling pests, diseases and weeds. In a system with reduced mechanical tillage based on mulch cover and biological tillage, alternatives have to be developed to control pests and weeds. Therefore, "Integrated Pest Management" becomes mandatory. One important element to achieve this is crop rotation, interrupting the infection chain between subsequent crops and making full use of the physical and chemical interactions between different plant species. Synthetic chemical pesticides, particularly herbicides, are in the first years inevitable but have to be used with very much care to reduce the negative impacts on soil life. To the extent that a new balance between the organisms of the farm-ecosystem, pests and beneficial organisms, crops and weeds, becomes established and the farmer learns to manage the cropping system, the use of synthetic pesticides and mineral fertilizer tends to decline to a level below the original "conventional" farming.

Therefore, although the entry point is a reduction of mechanical soil tillage, "Conservation Agriculture" (CA) involves a complete change in the crop production system. It involves modifications in the machinery, which means more mechanization, maintenance of surface residues providing at least 30 per cent soil cover, minimum soil disturbance, adjustment, if required, in the cropping system, minimum and need based use of chemicals [1].

8.1.1 Goal of Conservation Agriculture

Conservation Agriculture aims to conserve, improve and make more efficient use of natural resources like soil, water and biological resources combined with inputs. It contributes to environmental conservation as well as to enhanced and sustained agricultural production. It can also be referred to as resource-efficient/resource effective agriculture [2].

8.1.2 Characteristics of Conservation Agriculture

Conservation Agriculture maintains a permanent or semi-permanent organic soil cover. This can be a growing crop or dead mulch. Its function is to protect the soil physically from sun, rain and wind and to feed soil biota. The soil micro-organisms and soil fauna take over the tillage function and soil nutrient balancing. Mechanical tillage disturbs this process. Therefore, zero or minimum tillage and direct seeding are important elements of CA. A varied crop rotation is also important to avoid disease and pest problems. Rather than incorporating biomass such as green manure crops, cover crops or crop residues, in CA this is left on the soil surface. The dead biomass serves as physical protection of the soil surface and as substrate for the soil fauna. In this way mineralization is reduced and suitable soil levels of organic matter are built up and maintained.

8.1.3 What is not Conservation Agriculture

Zero-tillage: zero tillage is a technical component used in Conservation Agriculture but not everyone carrying out zero tillage is using Conservation Agriculture. Conservation agriculture not only avoids tillage by forcing the seed with appropriate direct drills into the soil, by maintaining a soil cover it also improves the structure of the soil. This facilitates direct planting. Conservation Agriculture uses biological tillage. Zero tillage as stand alone technique can also be applied in conventional agriculture under certain circumstances.

8.1.4 Conservation Tillage

Conservation tillage is practices that leave crop residues on the surface, which increases water infiltration and reduces erosion. It is a practice used in conventional agriculture to reduce the effects of tillage on soil erosion. However, it still depends on tillage as the structure-forming element in the soil. Nevertheless, conservation tillage practices such as zero tillage practices can be transition steps towards Conservation Agriculture.

8.1.5 Direct Planting/Seeding

This is only a technique that refers to seeding/planting without preparing a proper seedbed. The same equipment is used in Conservation Agriculture. However, the term direct seeding can also be used for implements, which combine primary and secondary tillage and seeding in one machine/tractor operation like the rotary till drills.

8.1.6 Organic Farming

Conservation Agriculture is not a synonym of organic farming, although it is based on natural processes. CA does not prohibit the use of farm chemical inputs. For example, herbicides are an important component in conservation agriculture, particularly in the transition phase, until the new balance of weed populations is managed. However, in view of the importance of the soil life for the system, farm chemicals, including fertilizer, are carefully applied and over the years, quantities applied tend to decline. In some cases organic farming can be practised within the CA framework [3].

8.1.7 Is CA Compatible with IPM

Conservation Agriculture is not only compatible but actually works on IPM principle. CA, like IPM, enhances biological processes. It expands the IPM practices from crop and pest management to land husbandry. Without the use of IPM practices the build up of soil biota for the biological tillage would not be possible.

8.1.8 What is the role of Animal Husbandry in Conservation Agriculture

Livestock production can be fully integrated into conservation agriculture, by making use of the recycling of nutrients. This reduces the environmental problems caused by concentrated intensive livestock production. Integration of livestock into agricultural production enables the farmer to introduce forage crops into the crop rotation, thus widening it and reducing pest problems. Forage crops can often be used as dual-purpose crops for fodder and soil cover. Particularly in arid areas with low production of biomass, the conflicts between the use of organic matter to feed the animals or to cover the soil has still to be resolved.

8.1.9 What are the Downsides of Conservation Agriculture

Conservation Agriculture may require the application of herbicides in the case of heavy weed infestation. During the transition phase, certain soil borne pests or pathogens might create new problems due to the change in the biological equilibrium. Once the Conservation Agriculture environment has stabilized it tends to be more stable than conventional agriculture. So far, there has been no pest problem that could not be overcome in Conservation Agriculture.

8.1.10 Benefits of Conservation Agriculture

Conservation agriculture attracts different people for different reasons:

8.1.10.1 Farmers level

- ☆ Reduction in labour, time, farm power.
- ☆ Reduction in cost.
- ☆ In case of mechanized farmers: longer lifetime and less repair of tractors, less power and fewer passes, hence much lower fuel consumption.
- ☆ More stable yields, particularly in dry years.
- ☆ Better traffic ability in the field.
- ☆ Gradually increasing yields with decreasing inputs.
- ☆ Increased profit, in some cases from the beginning, in all cases after a few years.

8.1.10.2 Communities Level/Environment Watershed

- ☆ More constant water flows in the rivers, re-emergence of dried wells. Cleaner water due to less erosion.
- ☆ Less flooding.
- ☆ Less impact of extreme climatic situations (hurricanes, drought etc.). Less cost for road and waterway maintenance.
- ☆ Better food security.

8.1.10.3 At Global Level

- ☆ Carbon sequestration (greenhouse effect): in some places no-till farmers start to receive carbon-grant payments; the global potential of Conservation Agriculture in carbon sequestration could equal the human made increase in CO_2 in the atmosphere.
- ☆ Less leaching of soil nutrients or chemicals into the ground water.
- ☆ Less pollution of the water.
- ☆ Practically no erosion (erosion is less than soil build up).
- ☆ Recharge of the aquifers through better infiltration.
- ☆ Less fuel use in agriculture.

8.1.11 Issues in Conservation Agriculture

Despite its advantages, CA has so far spread relatively slowly for a number of reasons. Firstly, there is greater pressure to adopt

Conservation Agriculture in tropical, rather than temperate climates. It has taken a long time, but over the past 20 years the establishment of a local knowledge base has ensured its spread. In some states of Brazil it is official policy, in Costa Rica the Ministry of Agriculture has a Department for Conservation Agriculture–so in these cases the policy makers have been convinced. The adoption of CA in the US was probably due to a mixture of public pressure to fight erosion and the financial incentives of reduced tillage. Europe is slowly getting there–farmers still do not feel sufficient pressure and environmental indicators (erosion, flooding) are not yet taken seriously enough. In India, farmers are moving towards zero tillage/conservation agriculture mainly due to increased costs of production especially that of fuel for tractors and other agricultural inputs and to some extent to increasing problem of water availability due to falling ground water table [1].

CA has great potential in Africa due to its propensity to control erosion, give more stable yields and reduce labour. There are a number of ongoing initiatives promoting different practices, from conservation tillage up to Conservation Agriculture. Another vast area where the adoption of CA would be extremely beneficial is Central Asia. In the countries of the former USSR conventional agriculture is virtually impossible because of environmental problems (erosion) and because of a lack of farm machinery, which has to be replaced. Unless Conservation Agriculture is adopted, the investment in new machinery will have to be very high. India can also be immensely benefited by adopting conservation agriculture by reducing the cost of production and be more competitive in the global market. Converting to Conservation Agriculture needs higher management skills, the first years might be very difficult for the farmer, therefore she/he might need moral support (from other farmers or from extension services) and perhaps even financial support (to invest into new machinery like zero-tillage planters). As it requires a complete change of understanding, the scientific and technical sectors often do not support Conservation Agriculture, fearing that they would contradict themselves.

8.1.12 Necessary Technologies are Often Unavailable

In order to try CA, the minimum a farmer needs is a zero tillage planter, which might not be available in the neighbourhood. Buying one without knowing the system or even having seen it, is a risk that few farmers take. Machinery dealers might not wish to promote CA as long as it is not supported by extension. This is partly due to the cost of the equipment but more importantly because the widespread adoption of CA will reduce machinery sales, particularly of large tractors.

8.1.13 Is Conservation Agriculture Real

Conservation Agriculture is being used on more than 45 M ha, mostly in South and North America. Its use is growing exponentially on small and large farms in South America, due to economic and environmental pressures. Farmers using CA in South America are highly organized (in regional, national and local farmers organizations), and are supported by institutions from North and South America. In Europe the European Conservation Agricultural Federation, a regional lobby group, has been founded. This body unites national CA associations in the UK, France, Germany, Italy, Portugal and Spain. In India also, during the last ten years zero tillage has increased from nil to more than 2 M ha and is spreading very fast. Now the emphasis is being shifted from conservation/ zero tillage to conservation agriculture and a few machines are in the testing stage that can seed into loose residues left after combine harvesting of the crop.

8.2 New Machines for Conservation Agriculture

Efforts are being made to develop and fine tune suitable machines for seeding into loose residues left after combine harvesting. The need is felt due to depletion of soil organic carbon and causing environmental pollution due to burning of crop residue after combine harvesting. At present four machines are under testing and evaluation for seeding direct seeded rice and wheat, which are briefly discussed hereunder [4].

8.2.1 Double Disc Coulters

This is one of the second-generation machines being tried under loose residue conditions. It has double disc coulters fitted in place of tynes to place the seed and fertiliser into the loose residues. The problem being faced with this machine is that being lightweight it fails to cut through the loose residues and the seed and fertiliser is dropped on the top of it, part of which reaches the soil surface. Irrigation is required immediately after seeding in order to facilitate the germination. This machine may work up to a residue load of about 4 to 5 U ha.

8.2.2 Punch Planter/Star Wheel

This is another machine, which is being tested for seeding into the loose residues. This mechanism is being used widely around the world under non-rice situations but its utility under rice wheat system with a residue load of 6 to 10 U ha is still to be proved. The initial results indicate that it may work under low residue load of up to 3 U ha. At present this machine drops the fertiliser on the surface in front of the moving star wheels, which is not the proper method of placing the fertiliser.

8.2.3 Happy Seeder

This is another machine, which cut and lift the residue in front, place seed and fertiliser using zero till machine and drop the chopped straw behind on to the seeded area. This machine is capable of seeding into the loose residue load of up to 10 U ha. Recently an improved version of this machine has been developed which cuts in strips of 5 cm only in front of tynes, which reduces the energy requirement of the machine.

8.2.4 Rotary Disk Drill (RDD)

This machine is based on the rotary till mechanism. The rotar is a horizontal transverse shaft having six to nine flanges fitted with straight discs for cutting effect similar to the wooden saw while rotating at 220 RPM. The rotary disc drill is mounted on the three point linkage system and is powered through the power take-off (PTO) shaft of tractor. The rotating discs cut the residue and simultaneously make a narrow slit into the soil to facilitate placement of seed and fertilizer. The machine can be used for seeding under conditions of loose residues as well as anchored and residue free conditions. If the machine is to be used under loose residue condition, it is better to use an offset double disc assembly for placement of seed and fertilizer otherwise inverted T-type or chisel type openers can also be used. The rotary disk drill can also be easily converted into rotary till drill by replacing the discs with L or J-shaped blades on the rotor. The rotor completely pulverizes the soil leading to a clean and fine tilth. In case rotary disk drill is to be used as a zero till drill, straight blades or discs can be used for minimal soil disturbance.

However, it must be remembered that in presence of loose residues, combination of rotary disk with coulter double disk completely avoids the raking problem of residues during seeding operations. Thus the newly designed rotary disk drill is a multipurpose machine, which can be used to seed under diverse situations depending upon the presence and condition of crop residues. The rotary disk drill can be used in manually as well as combine harvested fields for direct drilling of seed and fertiliser in a single tractor operation under variable field soil moisture conditions. Extensive field trials of the newly designed rotary disk drill are already underway to evaluate the performance and durability of rotary disks. The direct seeded rice and wheat crops were successfully established in loose residues up to 8 tonnes per hectare using rotary disc drill. The machine was also tested at the farmers' field for its capability to sow under zero till as well as under loose residue conditions and the result was very encouraging. However, one problem being faced was frequent blunting of the powered discs.

8.3 RCTs in Reclaimed Alkali Soils

Declining ground water levels and deteriorating quality of soil in the Indo-Gangetic plains is emerging a serious concern for agricultural sustainability in the near future. Depletion of groundwater in areas irrigated by tube wells and associated water quality concerns have brought about heightened awareness of the need for the judicious use of rain, surface and ground water resources[4]. The pressures on natural resources are immense. Soils are depleting in their fertility as a result of continuous and intensive cropping. The organic carbon levels in Punjab soils have decreased from 0.5 per cent (1950-60) to about 0.25 per cent at present. Tillage costs are rising, which accentuates the already serious labour shortages during peak periods of land preparation and harvest. For these and other reasons, the long-term sustainability of these systems is now a subject of attention. Further, increased cost of cultivation and declining productivity is compelling the farmers to quit the farming [4]. There is a general consensus that quality of natural resources base needs to be improved for enhanced productivity, sustainability and profit. Also, it is believed that future productivity growth would come through efficient management of inputs (water, nutrients and energy) and better risk management strategies. Targeted resource conserving technologies offer newer opportunities for better livelihood for the resource poor, small and marginal farmers of the region.

Though rice-wheat systems are critical in Indo-Gangetic plains, yet valuable information has remained underutilized by the farmers. Improved tillage, residue management and crop establishment practices show real potential for improving the productivity and profitability of rice-wheat systems. Reduced and zero tillage can improve yields, increase input-use-efficiency, reduced the intensity of machinery use and lower the production costs. Therefore, efficient soil and water management practices such as, tillage, irrigation and nutrients have to be fine-tuned according to the crop establishment requirements in a particular transect. Keeping the above facts in view, a long-term experiment is started at CSSRI to address some of the problems faced by the farmers in managing the natural resources for increasing the rice-wheat productivity.

8.3.1 Immediate Objectives

- ☆ To study the effect of resource conservation options on water, nutrient, energy-use-efficiency and crop productivity.
- ☆ To monitor the changes in physical, chemical and biological properties of soil as influenced by different resource conservation options.
- ☆ To work out salt, water, nutrient, energy and gas fluxes in selected treatments to monitor impacts of conservation agricultural practices in moderating climatic variations.
- ☆ To monitor changes in plant growth, yield and economics in different treatments.

8.3.2 Long-term Objectives

- ☆ To evaluate different resource conservation options for the sustainability of rice-wheat cropping system.
- ☆ To recommend the most profitable, eco-friendly and resource upgrading technology to the farmers of the region.
- ☆ To provide input for making suitable policy for the management of natural resources for rice-wheat cultivation in Indo-Gangetic Plains.

8.4 Resource Conserving Technologies–Potential Tools for Attaining Food, Nutritional and Livelihood Security

In early seventies, an estimated 920 million people in the developing world were chronically undernourished, with insufficient food but at the beginning of nineties, despite continuing population growth, the figure had been reduced to 840 million about 20 per cent of the total population of the developing world. Out of these, a major part lives in developing and poor countries in Asian and Africa continents. During the World Food Summit of 2002, the world leaders vowed to reduce this number to half by 2015. The UN Millennium Development Goals called for ensuring food as well as environmental security for all and can be achieved by increasing the crop productivity per unit of inputs (seed, fertilizer, water and land etc.) and there is an urgent need to arrange a square meal for every one without deteriorating natural resource base. But this will require a huge investment (public as well as private) in agriculture and allied sectors which are critical factor in achieving the goal. In view of this, top most priority should be given for investment in agriculture[5]. The International Food Policy Research Institute (IFPRI) estimated that world food production will grow by an average of 1.5 per cent per year between 1990 and 2020, if investments-such as in agricultural research, infrastructure, irrigation, markets and extension and training are maintained at least at 1980's levels. In India, agricultural and allied sectors contributes 25 per cent of GDP which affects the overall GDP by 0.5-1.0 per cent. As per our National Agriculture Policy, we should achieve 4.0 per cent growth rate but it is hovering around 1.50-1.75 per cent, which cannot ensure the food security of the people[4].

Tremendous changes have occurred in the world economy over the past two decades, but external development assistance has declined and agriculture has been disproportionately affected. Total external commitments to agriculture in 1994 were 23 per cent below those of 1980. Food aid has also declined from almost 17 million tonnes (cereal equivalent) in 1992-93 to around 9 million tonnes in 1994-95. For example, the investment in irrigation has slowed, especially in Asia. In India, since the mid 1990's private investment in agriculture has stagnated while public investment has continued to decline. It is also essential to increase long-term public investment in agriculture, roads, telecommunications, electricity and irrigation, which will stimulate private investment and contribute to a revial of growth momentum in agriculture. Massive donors backing is required for enhancing the investment in agricultural development, research, education and extension activities. Presently World Bank, International Monetary Fund (IMF), FAO, International Fund for Agricultural Development (IFAD), Asian Development Bank (ADB), USAID, and Department for International Development (DFID) are generally providing support to different projects in raising the status of food security of the people for developing and least developed countries. In such a situation, biotechnology, post harvest technology research and peri-urban agriculture should be given priority to increase production to feed the growing population within the shortest possible time.

Crop diversification, livestock raising, dairy production, and fisheries cultivation, these are all enthusiastic effect to built a reservoir of food for future food security of the population of the country. A carefully plan peri-urban agriculture had the potential to put agriculture on the path of self-propelled and self-reliant development under leadership through an indigenous public private partnership in the areas of food security and poverty [5]. Farm mechanization through resource conserving technologies (RCTs), *viz.* zero tillage, bed planting, mechanical rice transplanting and laser leveling are some of the efficient tools for increasing the crops production and their productivity and reducing the cost of production. In South Asia alone, there is about 14.0 million ha land which remains fallow after rice crop (rice fallows), mostly due to late harvesting of rice crop, poor irrigation facilities, excessive wetness and poor drainage. If these lands can be brought under cultivation, by adopting these technologies, to some extent can solve the food, nutritional and livelihood security problem of Asian countries and also generate additional employment opportunities for millions of inhabitants of this region. A major part of this falls in India alone (81.5 per cent), out of which about 50 per cent falls in Eastern IGP region. Crop establishment methods are exceedingly important factor in improving the productivity of crops. Land preparation cost represents a major portion of cost and with ever increasing oil prices, there is an urgent need to reduce the tillage practices for reducing the same and also the cost of fertilizers and water [5].

8.5 RCTs in Rice-Wheat System and Need for Conservation Agriculture

Earlier, Agriculture was focused on achieving food security through increased coverage under high yielding varieties, expansion of irrigation and increased use of external inputs. This enabled the rice-wheat (RW) to emerge as a major cropping system in the Indo-Gangetic plains (IGP) leading to the Green Revolution. These two crops together contribute more than 70 per cent to the total cereal production in India. The estimated area of RW system in India is around 10.0 M ha. At present, the food situation is comfortable but increasing production to meet the needs of ever growing population is full of uncertainties. With the increase in population, more and more land will be required for urbanization and productivity needs to be increased to meet the domestic and industrial demand [5].

The indiscriminate use, rather misuse, of natural resources especially water has led to the groundwater pollution as well as depletion of ground water resources. Presently we are sitting on the volcano and if the situation is not improved we will face water wars in the near future, the sign of which are quite visible in the surface water dispute between Punjab and Haryana and some other states in India. Depleting soil organic carbon status, decreasing soil fertility and reduced factor productivity are other issues of concern. These evidences indicate that rice-wheat system has weakened the natural resource base. If we continue to exploit the natural resources, the productivity and sustainability is bound to suffer. Therefore, in order to meet the aim of sustainable yields over time it is the need of the hour to avoid further degradation of the natural resources. Moreover, in the face of World Trade Organisation (WTO) regime, we must produce at lower cost to be competitive in the international market being India already a surplus state in food-grain production. To meet these needs, the agricultural system must develop cost effective technologies suitable for harnessing the untapped potential especially in the northeastern parts of the IGP (CASA, 2004). The resource conservation technologies to economies on cost of production and need for conservation agriculture are briefly discussed hereunder.

8.6 Laser Land Leveling

Laser land leveling is the process of smoothening the land surface within 2 cm from the average elevation of the field using laser equipped bucket which scraps from higher places and spread onto the low lying areas. This technology is a prerequisite for enhancing the benefits of other resource conservation technologies. Generally, fields are not properly leveled leading to poor performance of the crop, because, part of area suffers due to water stress and part due to excess of water. After laser leveling the field, it has been observed that yield enhances from 10 to 25 per cent. The higher yields are due to uniform crop stand, water distribution, crop growth and maturity. In addition to higher yield, this technology saves 35-45 per cent water, a scarce resource, due to higher application efficiency, increases nutrient use efficiency by 15-25 per cent, reduces weed problem and increases the cultivable area by 3 to 6 per cent due to reduction in area required for bunds and channels.

8.7 Resource Conservation Technologies in Wheat

8.7.1 Zero Tillage

This is a resource conservation technology in which wheat is directly seeded into the undisturbed soil after rice harvesting using a specially designed machine. In this seed and fertiliser is placed into narrow slits created by the knife type furrow openers of zero tillage ferti-seed drill. This technique was first adopted in the high yielding, more mechanized areas of northwestern India and Pakistan where lot of money was being invested on field preparation. This technology provided an opportunity to reduce the cost of cultivation by Rs 2500-3000 per hectare thereby increasing the profit margin of the farmers. In addition, development of resistance against commonly used herbicide 'isoproturon' in Phalaris minor was also responsible for its adoption in rice-wheat system due to lower incidence of this weed under zero tillage[6]. The work on zero tillage was initiated with the development of zero tillage machine by Dr. Bachan Singh and his group during 1992-93. The first version received at DWR had problem of patchy germination due to rigid side drive wheel, which slipped, at places near the bunds as well as where depressions were made due to tractor or combine harvester tracks and even human foot prints. Replacing the side drive wheel with balancing wheel and bringing the drive wheel to front and making it floating type solved this problem. The area under zero tillage was negligible till 1996-97 which increased slowly to 0.2 M ha during 2001-02. Thereafter the increase was very fast with less than 0.5 M ha in

2002-03 to around 2.0 M ha during 2004-05, including the area under reduced tillage. This speed of adoption was possible only because of farmer's participatory approach adopted by the scientists working on this technology.

8.7.2 Energy and Economics

Considering the total cost of field preparation and drilling or broadcast sowing of wheat, it was found that the maximum cost was Rs 1637 for broadcast sown wheat followed by drill sown (Rs 1413) and minimum was for zero tillage which was only Rs 179/ha. The total energy required for various tillage options varied from 20279 MJ/ha for zero tillage to 23631 MJ/ha for broadcast sown wheat. The benefit cost ratio was highest for zero and lowest for broadcast sown wheat, whereas, the specific energy (energy spent per kg of biomass production) requirement was lowest for zero and highest for broadcast sown wheat after conventional field preparation.

8.7.3 Reduced/Minimum Tillage

The impact of zero tillage is that most of the farmers have shifted from intensive tillage undertaking 6 to 12 tractor operations to reduced tillage involving 2 to 3 operations with various farm implements. Reduced tillage has advantage over conventional tillage as it saves on tillage cost with similar crop productivity. Reduced tillage has no apparent advantage over ZT but farmers are sometime forced to undertake 2 to 3 tractor operations due to the following reasons:

- ☆ With some of the farmers it a problem of mind set as they think that fields do not look tidy in the initial stages.
- ☆ Early transplanting or using early maturing varieties of rice vacates the fields by 15 t week of October whereas sowing of wheat is generally done from 15t week of November. In such fields standing stubble of rice become loose which creates problem in smooth running of ZT machine.
- ☆ Some broad leaved weeds like Rumex spp. germinate in the month of October after harvest of rice. Under such situations spray of Glyphosate is recommended in Zero tillage sowing of wheat. But it costs higher than 1-2

ploughing & also due to lack of proper spray technique some weeds are left after spray. Therefore, farmers prefer 1-2 tillage instead of spraying.

☆ The field becomes uneven due to formation of tracks in wet soil after combine harvesting. This happens either due to late irrigation or if rains occurs toward the end of rice season as it happened during 2004. Under such situation it becomes necessary to level those patches with 1 or 2 harrowing.

8.7.4 Rotary Tillage

The machine is a combination of rotavator, seed-cum fertiliser drill and a light planker-cum-driving wheel at the back. The machine uses a nine-row standard seed-cum-fertiliser drill, in addition to a rotary tiller. As the name of the machine suggests, it tills the soil, rather completely pulverises the top 8 to 10 cm, before placing seeds and fertilisers. The rotary tiller is a horizontal transverse shaft fitted with L-shaped blades. It is powered through the power take-off (PTO) shaft fitted at the rear of every tractor. The rotor completely pulverise the soil along with the cutting and mixing of residue or weeds present in the field leading to a clean very fine tilth. The machine has 7 gangs of six blades each and can be operated by a tractor of about 40 hp or more in dry/optimum moisture condition or about 30 hp or more for puddling of rice fields. The rotary-till drill can be used in manually harvested rice field at optimum soil moisture conditions for directly seeding wheat and fertiliser placement in a single tractor operation. This saves more than 70 per cent of the time and energy compared to conventional field preparation being followed by the farmers with 7 to 10 per cent higher yield. In addition, it helps in advancing the sowing in case the wheat seeding is getting delayed beyond 25th November. In combine harvested rice fields the machine can be used after removal or burning of the loose paddy straw. The field preparation by rotary tillage is better than conventional as the soil is completely pulverised. Incorporation of green manure crops, weeds and crop residues can also be accomplished in a single operation. As the seeding is done simultaneously, the soil moisture is conserved leading to better crop stand in addition to savings on energy, time and labour requirement making this technology more economic and eco-friendly. The machine was initially aimed to sow wheat directly after paddy harvesting. Subsequently, it was also tried for puddling of rice field since the rotary tillers are known to be excellent puddlers. Seed hoppers and seed placing units can be detached from the frame fixed on the rotary unit and the rotary unit alone can be used for puddling. A single operation after ponding of water was found sufficient for puddling followed by rice transplanting after 1 to 3 days depending upon soil type.

8.7.5 Bed Planting

This is a water saving technology, which saves 30 to 40 per cent water for growing wheat depending upon the soil type. In addition to water saving this technology has numerous advantages in rice-wheat system. Although there is no saving on the cost of land preparation or time but it can become cost effective by using the same beds for rice without reshaping. In this technology after preparation of land all three activities named bed formation, placement of fertilizer and sowing of wheat are done in single operation. Crop cultivars are known to vary significantly in their performance on raised beds. It is suitable for seed production also because of bolder grain and easier rouging. It reduces the herbicide dependence due to mechanical weed control with the same bed planter fitted with inter culture tines with simultaneous placement of fertiliser. In situations where sowing can be delayed due to pre-sowing irrigation dry seeding can be done on raised beds followed by irrigation immediately after seeding. Irrigation can also be given at grain filling stage, which is generally avoided by the farmers for fear of crop lodging. In this technology nitrogen use efficiency is also higher because of light irrigation and top dressing on beds.

8.7.6 Surface Seeding

This is a technology, which does not require any field preparation and sowing of wheat is done in standing rice crop a few days before or immediately after rice harvesting. There are areas in the eastern part of Indo-Gangetic plains where land remains wet after rice harvesting for a long time and field preparation for sowing second crop is not possible. Under such conditions surface seeding provides an opportunity to take wheat crop in rabi season. Even in areas where field preparation is possible wheat sowing is delayed leading to very low yields. So by adopting surface seeding one can harvest higher yields. In this technology dry or soaked seed is broadcasted over the wet soil. To prevent bird damage the seed is invariably coated with cow dung. For proper and uniform crop stand drum seeders can also be usefully employed after rice harvesting. Nowadays, farmers are practicing the surface seeding successfully not only in wheat but also in other upland crops like pea, gram, lentil etc.

8.7.7 Water Use and Savings Under Various Tillage Options

There are two school of thoughts among the scientists, one who says that zero tillage can save up to 25 per cent water whereas the other group feels that such a huge savings is not possible although there may be some savings on account of comparatively lower infiltration rate and subsequently higher retention of moisture in zero till sown fields. The argument the first group put forward is that wheat can be sown without pre-sowing irrigation. This argument doesn't carry much weight because the soil can be prepared using the same moisture and wheat can be sown almost simultaneously even under conventional field preparation. If the scenario of rice residue incorporation followed by heavy irrigation of about 10 cm to decompose it is considered, then there is possibility of water saving up to 20-25 per cent. To address this issue a study was conducted during *rabi* 2001-02 season involving wheat sowing under three tillage options (ZT, FIRBS and CT) in manually harvested (near to ground) rice field. A pre-germination irrigation was applied after 3 days of seeding in Furrow Irrigated Raised Bed-planting (FIRB) system for ensuring germination and good crop stand. Thereafter, need-based irrigations were applied. The water applied at each irrigation was measured using a Parshal flume. At each irrigation, water applied was marginally lower under zero tillage whereas in FIRBS it was only around 35 to 40 per cent compared to conventional tillage. The total post seeding irrigation water requirement was about 25 cm in conventional tillage whereas zero tillage required about 3 per cent less irrigation water. The irrigation water saving was more than 30 per cent in case of FIRB system of wheat cultivation.

The average yield of wheat in various tillage options varied from 60.35 q ha^{-1} in FIRB system to 64.06 q ha^{-1} under zero tillage. The total water use calculated by accounting for the rainfall and profile depletion varied from around 30 cm in FIRBS to about 37 cm in conventional field preparation whereas zero tillage required about

36 cm of water. The total water saving compared to conventional was 2.58 per cent in zero tillage and about 20 per cent in FIRB system of wheat cultivation. The water use efficiency was highest in FIRBS and the lowest in conventional system of field preparation. It can therefore be summed up that there is no great water saving or usage efficiency under ZT even though numerically there is a minor advantage.

8.7.8 Eco-friendly Tillage Options

These alternate tillage technologies are also environmental friendly compared to traditional broadcast sown wheat. The carbon dioxide emission due to burning of fuel (assuming 2.6 kg CO_2 production/litre of diesel burnt) during field preparation was 208.00 kg/ha in conventional farmers' practice and was only 15.60 kg/ha in zero tillage and 36.92 kg/ha in rotary tillage. Even conventional field preparation followed by drill or bed planter sowing on FIRBS resulted in reduction of carbon dioxide emission by 18–19 per cent compared to broadcast sown wheat. This corresponding reduction in CO_2 emission comes to 82.25 and 92.50 per cent for rotary and zero tillage, respectively.

8.8 RCTs in Rice

Wet tillage *i.e.* puddling for growing rice has been suspected to adversely affect the soil and water resources and efforts are being focused on developing and fine tuning the technologies to grow direct seeded rice or transplanted rice without wet tillage. The initial results are encouraging and are discussed briefly here. Direct dry seeded and unpuddled transplanted rice: In this system rice is grown like any other upland crop with seed placed in the soil by seed cum fertilizer drill with or without ploughing. The traditional practice of growing rice *i.e.* transplanting in puddle conditions requires higher amount of water and labour. In puddle soil once the water dries, cracks develop and water percolates beyond the root zone along with nutrients. Generally, plant population is 18-20 in manual random transplanting against the recommended density of 35-40 plants, which is a major constraint in achieving higher yield of transplanted rice [6].

Direct seeding has advantages of faster and easier planting, reduced labour and less drudgery with earlier crop maturity by 7-10 days, more efficient water use and higher tolerance of water deficit, less methane emission and often higher profit in areas with an assured water supply. Thus the area under direct seeded rice has been increasing as farmers in Asia seek higher productivity and profitability to offset increasing costs and scarcity of farm labour. Weed control is a major issue in direct seeded rice and to over come this problem intensive efforts are being made by the agricultural scientists. In some soils, spray of micronutrient like Zn and iron may be needed to remove their deficiency. Direct seeding of rice using zero till drill, rotary till drill, drum seeder as well as broadcasting under various field preparation or puddling options was tried at DWR research farm. Seeding depth was kept at 2-3 cm while using drill for seeding. For comparison purposes transplanting was also done under conventional puddling as well as under zero tillage and after field preparation with rotary tiller. The rice variety used was IR 64. Direct seeding was done in the first week of June on the same day when nursery was sown for transplanting. For weed control Sofit @ 1500 ml/ha was applied after four days of direct seeding followed by one weeding at around 35 days after seeding.

Among the direct seeding options, the yield recorded was highest where rice was seeded using rotary till drill followed by broadcasting sprouted rice seed after field preparation by rotary tillage and lowest when broad casted under zero tillage. The mean yield in rotary tillage was significantly higher compared to zero tillage. Direct drilling by zero till drill and rotary till drill was at par and as good as transplanting under zero tillage or after field preparation by rotary tillage and was significantly higher than drum seeding or broadcasting under zero tillage. Among transplanting and direct seeding options, highest yield was recorded in machine transplanting, which was significantly better than broadcasting and drum seeder but statisticallly at par with other transplanting or seeding options. The yield was marginally higher in conventionally puddled conditions compared to transplanting without tillage, after field preparation by rotary tillage or direct drilling by zero or rotary till drill.

8.8.1 Direct Wet Seeded Rice

In this system sprouted seeds are broadcasted or placed with drum seeder under puddled or unpuddled conditions. Wet seeded rice also reduces labour costs and effective herbicides for weed control has helped making this technology more popular. Seed rate in drum seeded rice varies from 50-75 kg/ha whereas in broadcasting method of seeding 20-30 kg/ha is sufficient. In wet seeded rice puddling can be avoided with out any adverse effect on rice yield. The observations at farmers field showed that mortality of sprouted seeds are higher under puddled compared to unpuddled conditions.

A field trial on direct seeded rice was conducted with different seed rates varying from 30 to 80 kg/ha during 2002. Similar yield was recorded at varying seed rates suggesting that the seed rate can be further reduced. In 2003 rice season, an additional treatment of 20 kg/ha was included. The varying seed rates were kept based on earlier recommendation of the Directorate of Rice Research of 75-100 kg/ha. The variety used was IR 64 having a 1000 grain weight of about 26 grams. For a population of about 0.33×10^6 plants/ha recommended for transplanted rice, the seed requirement is likely to be around 11 kg/ha after giving an allowance of 20 per cent loss in germination percentage of seed. If rodent and bird damage are further added to the estimates, almost double the seed requirement (20 kg/ha) should be good enough. The trial was sown in the first week of July during 2002 and second week during 2003 when the transplanting is generally done. The yield recorded was almost similar at seed rates of 20 to 80 kg/ha.

8.8.2 Weed Management in Direct Seeded Rice

Weed management is the major problem in direct seeded rice (DSR). Experiments have sown that it can be tackled successfully by integrated weed management practices which include stale bed technique, crop rotation, brown manuring, zero tillage, use of competitive varieties, water management, mulching, intercropping of cover crops and use of suitable chemicals at right time. Integrated weed management approach utilizes all suitable techniques and methods, which maintains the weed population below economic threshold level [6].

8.8.3 Leaf Colour Chart

Leaf colour is a fairly good indicator of the nitrogen status of plant. Nitrogen use can be optimised by matching its supply to the crop demand as observed through change in the leaf chlorophyll content and leaf colour. The leaf colour chart developed by International Rice Research Institute, Philippines can help the farmers because the leaf colour intensity relates to leaf nitrogen status in rice plant. The monitoring of leaf colour using leaf colour chart helps in the determination of right time of nitrogen application. Use of leaf colour chart is simple, easy and cheap under all situations. The studies indicate that nitrogen can be saved from 10 to 15 per cent using the leaf colour chart.

8.9 References

1. CASA (2004) Groundwater use in northwest India culture for advancement of sustainable agriculture, New Delhi, India.
2. Chand, S., Trag A.R, Dar, N.A., Hasan, B. and George, E. (2010) Resource conservation technologies for sustainable agriculture. Book of Abstract, International conferences on soil fertility and soil productivity, Humbolt University, Berlin. Pp 228.
3. Chand, S. (2008) Integrated Nutrient Management for Sustaining Crop Productivity and Soil Health, I. B. D. Co. Lucknow. pp112.
4. Sharma, D.P., Sharma, S.K., Joshi, P.K., Singh, S. and Singh, G. (2008). Resource conservation technologies in reclaimed alkali soils. Technical Bulletin, I. Central Soil Salinity Research Institute Karnal, India.
5. Sharma, R.K. (2006) Sustainable irrigated agriculture through command area development. National level training course command area development with emphasis on land leveling shaping planning and design. Central Soil Salinity Research Institute Karnal. India. pp 153-167.
6. Tomar, R.K., Sahoo, R.N., Garg, R.N. and Gupta, V.K. (2006) Resource conserving technologies potential tools for attaining food, nutritional and livelihood Security. *Indian Farming*. pp. 24-31.

Chapter 9

Advances of Nutrient Management

"Integrated nutrient management is best way to enhance soil health and crop productivity."

—***Anonymous***

Agriculture is the main stay of the developing countries and Indian economy, contributing about 22 per cent of the gross domestic product (GDP) and providing livelihood to two thirds population. Application of all the needed nutrients through fertilizers had adverse effect on soil fertility leading to unsustainable production for longer time; while integration of chemical fertilizers with organic manures not only maintain soil fertility but sustain crop productivity also. An efficient nutrient management system would minimize loss of nutrients, saving unnecessary input cost. Efficient nutrient use is essentially an offspring of balanced nutrient use and sound management practices and decisions. Balanced nutrient use is not only the first requirement; it is rather a pre-requisite since no amount of agronomic manipulation can produce high efficiency out of an imbalanced nutrient dose. Crop productivity and soil fertility are interlinked. But ultimately it is improvement in the nutrient-use efficiency that will decide crop production, productivity per unit area and long term sustainability on term basis.

Keywords: *Bio-fertilisers, FYM, Organic Carbon, Soil Health.*

9.1 Brief Reviews

Integration of farmyard manure and organic sources exhibited an increase in yield and yield related attributes of crops. This could be due to balanced C/N ratio, more decomposition, more mineralization, more availability of native and applied macro and micro-nutrients. All these might have accelerated the synthesis of carbohydrates and its better translocation from sink to source that might have led to an improvement in yield and yield related attributes. [1]Gill *et al.,* have been conducted a long term experiment at several locations and find that the combined use of FYM or green manure @ 6 t/ha at the time of rice transplanting along with 50 per cent recommended NPK not only gave as good yield as compared with 100 per cent recommended NPK, but it saved 50 per cent expenditure on fertilizers and post harvest availability of soil organic carbon, N, P, K status of the soils also improved. An other experiment conducted by [2] Munda *et al.,* and concluded that the application of FYM 2.5 t/ha + *Eupatorium* 2.5 t/ha with 100 per cent RDF to rice + 50 per cent RDF to toria was beneficial in terms of rice equivalent yield and economics of rice-toria cropping system under mid hills dry terraces of Meghalaya. [3]Singh *et al.,* also concluded on the basis of two year experimentation that the growing of wheat after green gram or clusterbean (leguminous crop) with integrated application of FYM @7.5 t/ha + 50 per cent RDF (50 kg N + 13.25 kg P/ha) + Biofertilizers (Azotobactor + PSB) is effective and economically beneficial for attaining maximum and sustainable yield of wheat and crop sequence in the arid region of Rajasthan. Integration of FYM along with chemical nitrogen significantly increased the yield of wheat as well as of succeeding green gram[4]. Chand and Somani[5] while working in a field experiment on integrated nutrient management in mustard recorded a significant increased in seed as well as oil yield with the treatments receiving 50 per cent NPKS +FYM @ 10t/ha + co-inoculation of azotobactor +PSB. [6]Singh *et al.,* conducted an experiment during *kharif* 2005 to *rabi* 2006-07 at Modipuram, Meerut (UP). Results reveled that in rice and wheat 25 to 50 per cent NPK can be substituted by FYM, vermicompost or FYM + vermicompost. They also found that more removal of N, P and K in Chemically fertilized plots at similar yield level clearly indicates luxurious absorption of NPK due to quick availability of nutrients in comparison to integrated treatments. [7]Tiwari *et al.,* was recorded maximum grain yield of rice (3.69 t/ha) with 50 per cent recommended dose of nitrogen through inorganic fertilizer + BGA 10 kg/ha + FYM @ 5 t/ha. This may be due to slow and prolonged availability of nutrients to the plants. [8]Virdia *et al.*, also concluded that integrated nutrient management improve rice yield, as proper decomposition of organic matter supply available plant nutrient directly to plants and created favourable soil environment, ultimately increased the nutrient capacity of soil for longer time, which resulted in better growth, yield attributes and ultimately grain and straw yield. [9]Shivakumar and Ahlawat revealed that the combined application of 5 t/ha each crop residue and FYM and 5 kg/ha zinc along with 100 per cent RDF to soybean will be helpful in realizing higher productivity and net returns from both the crops individually as well as from soybean-wheat cropping system. These treatments also help in maintaining higher available status of N, P, K and Zn besides higher organic carbon content.

Qureshi *et al.*[10] observed that the highest seed and stover yield of Pea (*Pisum sativum*) 30 and 55 q/ha, respectively, was recorded with the application of 50 per cent NPKS + 20 t FYM/ha + *Rhizobium*. The treatment was significantly increased the available nutrient status over control. It may be due to the liberation of carbon dioxide and release of organic acids during mineralization of organic manure which increased the availability of nutrients and also fixes atmospheric nitrogen and phosphorus by secretion of organic acids and phosphate enzymes. Sole applications of organic or inorganic fertilizers are in no way a suitable solution for maintaining soil health and enhancing crop productivity. So the solution lies in the integrated use of chemical fertilizers and organic manures for obtaining sustainable crop production, better nutrient availability and efficient nutrient use, besides reducing nutrient losses [11] and improving fruit quality [12]. Due to prolonged cultivation of crops with recommended dose of inorganic fertilizers alone, the productivity of soil has gone down and time has come to supplement these inorganic fertilizers with organics and micronutrients to sustain the fertility and productivity of the soils. Since organics and inorganic alone can not meet the nutrient requirement, INM holds not only the great promise in crop production but also against emergence of multiple nutrient deficiencies and deterioration of soil health for livelihood security.

9.2 Introduction

The fertilizers are very important sources of plant nutrients and played a prominent role in increasing food grain production of the country. About fifty per cent increase in the food grain production in post green revolution era is attributed to the use of fertilizers. Soil is a precious natural resource equally as important as water and air. The proper use of soil greatly determines the capability of a life-support system and the socio-economic development of a country. The agriculture era has been changed from resource degrading to resource conserving technologies and practices which will enable help for increasing crop productivity besides maintaining soil health for future generations. The INM provides an excellent opportunity not only for sustaining soil but enhancing crop productivity also. The INM is the maintenance or adjustment of soil fertility and plant nutrient supply to an optimum level for sustaining the desired crop production through optimization of the benefits from all possible sources of plant nutrients in an integrated manner. INM options proved to be helpful in food, nutritional and livelihood security besides socio-economic conditions of Indian farmers.

To overcome the negative effects of application of plant nutrients, both at low or imbalanced and high levels of input can be avoided by excellent management. Balanced fertilization supplemented with organic nutrient sources help in overcoming the hazards of nutrient depletion and of mining soil fertility. Integrated Nutrient Management (INM) provides excellent opportunities to overcome all the imbalances besides sustaining soil health and enhancing crop production. The concept of INM is the maintenance or adjustment of soil fertility and of plant nutrient supply to an optimum level for sustaining the desired crop production through optimization of the benefits from all possible sources of plant nutrients in an integrated manner. The INM as defined by Harmsen[13] here differs from the conventional nutrient management by more explicitly considering nutrient from different sources, notably organic materials, nutrients carried over from previous cropping seasons, the dynamics, transformations and interactions of nutrients in soil, interaction between nutrients, their availability in the rooting zone and during growing season in relation to the nutrient demand by the crop. In addition it integrates the objectives of production, ecology, environmental and is an important part of any sustainable agricultural production system. The objectives is to maintain or enhance soil productivity through a balanced use of mineral fertilizers combined with organic and biological sources of plant nutrients.To improve the stock of plant nutrients in the soil. To improve the efficiency of plant nutrients, thus limiting losses to the environment. The effect on soil organic carbon after the application of different organic residues under different cropping sequences in Table 9.1.

Table 9.1: Effect on Soil Organic Carbon After the Application of Different Organic Residues Under Different Cropping Sequences [14].

Land Use	*Material Added*	*Organic Carbon (per cent)*
Maize-wheat (25 yrs)	Control	0.51
	FYM	2.49
Cotton-sorghum	Control	0.56
(45 yrs)	FYM	1.14
Ragi-cowpea-maize	Control	0.30
(3 yrs)	FYM	0.64
Rice-rice (10 yrs)	Control	0.43
	50 per cent from inorganic + 50 per cent through green manuring (*Sesbania aculeate*)	0.90
Rice-wheat (3 yrs)	Control	0.44
	FYM	0.54
Rice-wheat	Fellow	0.23
(7 yrs)	Green Manuring (*Sesbania aculeate*)	0.37

Intensive rice cropping with short-duration high-yielding varieties along with increased use of mineral fertilisers and improved irrigation facilities have resulted in spectacular increases in crop productivity. This has, however, led to gradual replacement of organic manures as sources of plant nutrients. There has been a sharp increase in the prices of P and K fertilisers following withdrawal of subsidy, which has led to their decreased consumption by the farmers. The low purchasing power of the farming community and the issue of soil health have again renewed interest in organic recycling. Organic sources available for use in rice production include the bulky organic manures like FYM, quick growing leguminous shrubs grown in the cropping sequence, leguminous trees grown in alley formations and using their loppings as mulch materials, forage or food legumes properly inoculated with Rhizobia and grown in the sequence, blue green algae and Azolla. Yield potential of rice can be realized in India only by maintaining a balance between supply and demand of nutrients by integration of inorganic and organic sources of nutrient like farmyard manure, press mud, crop residue, green manure brown manure, loppings, green leaves and twinges, vermicompost, sheep and goat manures, poultry manures, aquatic weed compost biofertilisers, azolla, blue green algae, animal urine etc.

Long-term experiments have shown that neither organic sources nor mineral fertilisers alone can achieve sustainability in crop production. Continuous use of FYM is effective in stabilizing rice productivity under low to medium cropping intensity where the nutrient demand is relatively small. Nonetheless, integrated use of organic and mineral fertilisers has been found to be more effective in maintaining higher productivity and stability through correction of deficiencies of secondary and micronutrients in the course of mineralization on one hand and favourable physical and soil ecological conditions on the other. Organic manuring also improves the physical and microbial conditions of soil and enhance fertiliser use efficiency when applied in conjunction with mineral fertilisers. Thus, all the major sources of plant nutrients such as soil, mineral, organic and biological should be utilised in an efficient and judicious manner for sustainable crop production of rice. Locally available organic materials such as chopped straw, FYM, water hyacinth compost, Azolla and green manure *in situ* with sunnhemp and dhaincha can substitute N fertilizer up to 50 per cent of the total crop requirement. The grain yield of kharif rice was increased by 21-22 per cent and 10-13 per cent with the application of Azolla or wheat straw and FYM or water hyacinth compost, respectively. The residual effect of these sources on the succeeding Rabi rice showed a yield increase of 14-18 per cent and 8-10 per cent compared with the control [15].

The yield potential of a crop will be limited by any nutrient, that soil cannot adequately supply. Poor crop response to one nutrient can often be linked to a deficiency in another nutrient or other management technique. Sometimes, poor crop response can also be linked to basic soil imbalances like acidity, sodicity or salinity, or a problem with beneficial soil micro-organisms which is responsible for inadequate supply of essential nutrients to crop plant. Although nitrogen is the most commonly discussed nutrient, many others are also essential for plant growth. Other once that are most common nutrient deficiencies after nitrogen (N) are phosphorus (P), zinc (Zn), and sulphur (S). Potassium (K) levels are usually adequate although deficiencies are sometimes experienced on cultivation. Deficiencies of boron (B) and copper (Cu) have also been recorded. Molybdenum (Mo) deficiency and manganese (Mn) toxicity can occur on the more acid soils. Thus the balance supply of nutrients as per deficiency in soil and requirement of crop in a efficient system is a noble strategy to reduce rice yield gap for longer time.

9.3 Organic Resources

Farm and animal wastes from cattle, sheep, goat, pig, poultry and fish in India have a nutrient potential of 7.2 million tones. Human excreta can add another 4.7 million tones. Organic manures contribute several nutrients in highly variable and low quantities depending on the kind of species, the quantity of the nourishment received by the manure generating plant and animal and the efficiency of the method of preparation. The potential of rural and urban compost in India is estimated to be 800 and 16 mt respectively. Less than 50 per cent of the manurial potential of the livestock population is utilized at present in crop production. The major contributor of rural compost is animal dung, which has a potential of about 7 mt of NPK. Night soil if properly exploited can provide about 5 mt of NPK nutrients. About 1/3 of the residue potential is available for utilization in agricultural production. About 100 mt of crop residues are produced in the country which have potential of supplying about 7.3 mt of NPK.

9.3.1 Incorporation of Green Manure in Rice Field

Legume grown as on intercrop do not compete for nitrogen with the component crop. On the contrary legume may provide some nitrogen benefit to associated crop. For example, Bandyopadhyay[16] and De used N^{15} data to show that sorghum derives part of nitrogen from the soil pool enriched by concurrently grown legumes. The practice of green manuring for improving soil fertility and supplying of nutrient requirement of crop is aged old. Green manure of legume shrubs or tree loppings has been known to be beneficial for sustaining rice productivity. Sunnhemp and dhaincha are popular legumes for green manuring in rice and can accumulate up to 100 kg N/ha in 50-55 days. Incorporation of these green manures in situ before transplanting rice supplies about 45-60 kg N/ha, besides providing a significant residual effect to the succeeding crops. Fertilizer use efficiency is improved when a legume crop such as *Sesbania cannabina* or *Lathyrus sativus* is introduced in rice based cropping system. Adding loppings of leguminous trees like *Leucaena leucocephalla* and *Glyricidia napus* grown in alleys or any other leguminous trees like *Robinia psudocasia, Dalbersia sisoo* etc. green leaves and twinges can also meet the crop N requirement substantially. The productivity of rice-based cropping system can be increased by about 1 t/ha besides a net saving of 30 kg fertilizer N/ha by including a short-duration legume such as cowpea or greengram and incorporating its residues into the soil after harvesting the grains. Similarly, blue green algae culture in the rice field can contribute about 25 kg N/ha to the rice crop. Algae multiply and cover the field like a carpet which when incorporated into the soil, decomposes and releases N for rice crop. Azolla can be grown in tanks or in rice fields and incorporated into the soil after 4-6 weeks. Nitrogen contribution through Azolla dual cropping with rice has been worked out to be about 25-30 kg N/ha. Azolla growth is generally poor without P fertilisation but a substantial improvement in growth and yield of rice is achieved when dual cropped Azolla is fertilized with P and incorporated into the soil at a later stage. An application of about 30 kg P_2O_5/ha is adequate for optimum growth of Azolla.

Table 9.2: Effect of Cowpea Green Manure on Rice and Wheat Grown in Sequence

*Crop and Fertilizer Rate***			*Yield (kg/ha)*		
Cowpea	*Rice*	*Wheat*	*Rice*	*Wheat*	*Total*
Fallow	120:30:30	120:60:30	6300	4765	11065
GM*	120:30:30	120:60:30	7040	5010	12050
GM	90:22:22	120:60:30	6834	4935	11769
GM	60:15:15	120:60:30	6460	4915	11375

GM*-Green manure, ** Rate of N: P_2O_5: K_2O/ha

9.3.2 Brown Manuring in Rice

Instead of traditional practice of growing green manuring crop and incorporating it in soil before rice transplanting which led to more water consumption and fuel cost, a new technique of co-culture is developed. This comprises of growing of both rice and *Sesbania* together and subsequently *Sesbania* crop is knocked down with 2, 4-D @ 500 g ha^{-1}. It also reduces weed Population by half.

Plate 9.1: Brown Manuring in Rice

9.3.3 Biofertilizers

These are low cost agricultural environment friendly input used as seed inoculation and also soil inoculation. Inoculation of Azotobacter and Azospirillum substitute, 22 and 20 kg N/ha, respectively. Blue green algae (BGA) applied @ 10 kg/ha gave a saving of 20-30 kg N/ha and Azolla @ 6-12 t/ha had an N equivalent of 3-4 kg/t. Mixed Biofertilizer formulations consisting of nitrogen fixing organisms and phosphate solubilizing bacteria (PSB) proved superior to individual inoculants. In wetland rice inoculation of 'Azophos' (*Azospirillum* and PSB) and PGPR (*Pseudomonas*) along with 75 per cent NPK gave maximum rice grain yield (6250 kg/ha), an increase of 5.8 per cent over 100 per cent NPK alone (5905 kg/ha) at Coimbatore. In on-farm trials on large field plots on rice at Amaravathi, Andhra Pradesh, dual inoculation of *Azospirillum* and PSB at 100 per cent N, gave 5.35 tonnes/ha yield, whereas 100 per cent N alone yielded 4.77 tonnes/ha. The 75 per cent N + dual inoculation gave 5.03 tonnes/ha and, thus, saved 25 per cent nitrogen.

9.3.4 Crop Residues

A large proportion of the crop residues in India is used as animal feed and fuel in the rural homes. Some residues also have industrial uses. The straw obtained from wheat, maize, grain sorghum and pearl millet is value more as an animal feed. Legume residues are widely used as feed because of their high protein content, reducing their availability for recycling. In the rice-wheat cropping system itself, each crop under good management produces 5-6 t/ha of straw. Most of the paddy straw on the large farms is being burnt *in situ*. Wheat residue left after combining on large farms is also being burnt. One of the reasons of burning straw is the obstructions it creates in tillage operations. Total crop residue in India is estimated to be around 400 million tones. Assuming that even one-third of it is available for recycling, the primary nutrient contained in them taking the mean concentration in rice and wheat straw (0.55 per cent N + 0.17 per cent P_2O_5 + 1.28 per cent K_2O) works out to 2 million tones and assuming that 50 per cent of it could be mineralized annually, the quantum of primary nutrients made available to the crops in a year would be about one million tonne besides micro-nutrients in variable quantities added to the soil.

9.4 Management of Fertilizers

The fertilizers have played a prominent role in increasing food grain production of country. About 50 per cent increase in agricultural production in post green revolution era is attributed to the use of fertilizers. The use of chemical fertilizer would remain the mainstay of agricultural production in future as well, given the increasing food demands of growing population and insufficient availability of alternative nutrient sources. The country will require about 300mt of food grains by 2025 to feed around 1.4 billion populations. This would use of about 45 mt. of nutrients[17].

9.4.1 Nitrogen Fertilizers

Use of nitrification inhibitors and slow release N fertilizers can increase the efficiency of applied fertilizer N and this led to the discovery of nitrification inhibiting capacity of neem. The development of coating urea with neem oil micro-emulsion and suitable equipment has made it possible to take this technology to the industry. NFL has already adopted the technology on a fairly large scale and other fertilizer manufacturers such as Indo-Gulf Fertilisers Ltd. (at Jagdishpur) and Shriram Fertilisers and Chemicals (at Kota) are on way to produce neem coated urea using neem oil micro-emulsion technology. Neem cake coated urea can increase the efficiency of urea at least by 5-10 per cent, resulting in 5-10 per cent increase in yield with the same amount of fertilizers or in a saving of 15-20 per cent N at the same yield level. Placement of fertilizer N is equally important to prevent ammonia volatilization losses, which could easily by 5-15 per cent. This is important at the national level. Assuming that about half of the fertilizer N is applied at sowing when its placement is possible and taking the consumption of N at 11 million tones per year, loss of about 0.25-0.5 million tones of N (equivalent of the productivity of about 2 urea plants per year) by ammonia volatilization can be prevented [18]. Creating the farmers awareness to such losses will go a long-way to reduce ammonia-volatilisation losses. Development of suitable equipment for small scale farmers for placement of N and P is highly desirable; large-scale farmers use fertilizer-cum-seed drill for most upland crops.

9.4.2 Phosphorus Management

Advantage of P placement is re-emphasised considering the fact that this is the costliest plant nutrient and needs to be most efficiently utilized. During 2000-02, 0.49 mt of P_2O_5 was imported in addition to 3.9 mt. production in India. Furthermore, bulk of rock phosphate, sulphur and phosphoric acid needed to produce phosphate fertilizer is imported. This makes it imperative to increase its use efficiency. Nearly 0.7 mt of ground rock phosphate is sold for direct application to soil and most of this is low grade (15-20 per cent P_2O_5 material). Efficiency of low-grade indigenous rock phosphate can be increased by a number of techniques. These include: (1) mixing it with soluble P fertilizer (II). Mixing it with elemental S, (III). Mixing it with iron pyrites, (IV). Use of phosphate solubilising organism (PSB) along with crop residue incorporation.

The last technique *i.e.* combined use of PSB and cereal residue incorporation again brings out the advantage of incorporation of rice/wheat residue in rice-wheat cropping system.

9.4.3 Potassium Management

Potassium is becoming a major bottleneck in Indian agriculture because of its scarcity of natural deposits in the country and prices hike due to removal of subsidies. The country has immense potential of organic manures, crop residues, and similar recyclable wastes which may scientifically be processed for beneficiation and applied for promoting soil health and crop yields.

9.4.4 Secondary and Micronutrients

The gains achieved so far in balanced fertilizer (NPK) use for crop production have been distorted due to the fertilizer decontrol policy making P and K nutrients costly. This has promoted high N usage (64 per cent of the total fertilizer consumption) that causes more mining of P, K: secondary and micronutrients. Even many farmers use diammonium phosphate (DAP) as a source of N and P fertilizer particularly in oilseeds and pulses because of its easy application and availability. But this leads to non-addition of essential elemental suphur for oilseeds. With wide spread incidence of deficiencies of secondary and micronutrients it becomes practically impossible to increase the crop response to fertilizer unless adequate replenishment of these deficient nutrients is ensured. At Ludhiana, seed yield of hybrid sunflower increased from 2.1 t ha^{-1} in control to 3.2 t ha^{-1} with the application of NPKS. Under Ca deficient conditions, application of Ca through gypsum improved the uptake of P and K in groundnut besides Ca. Red and laterite soils of Jharkhand, Assam, Orissa and West Bengal are deficient in available B where application of 5 to 20 kg borax ha^{-1} along with NPK produced rice grain yield response ranging from 0.13 to 1.5 t ha^{-1} and wheat grain yield response from 9.17 to 0.57 t ha^{-1}. Spectacular response of oilseeds and pulses to soil application of B (1.0 to 6.0 kg ha^{-1}) has also been observed in these B deficient soils [19]. Swell shrink soils are inherently low in micronutrient particularly Fe and B there was a significant increase in grain and straw yield of rice with increase in both NPK and Fe level and the beneficial effect of applied Fe on crop yield is likely to be more at high NPK.

Boron is a micronutrient of special importance because of its role in the fertilization and flowering processes of crops. Due to the vital role it plays in these processes, if it is deficient, one of the first adverse affects is on flowering and fruiting and therefore on the yield and/or quality of the seeds and fruits. Application of boron as directed spray @ 0.2 per cent at the time of ray floret initiation stage improves the seed yield of sunflower up to 24 per cent due to the dual role of increased seed set and seed filling, there by increasing test weight and seed yield. Among micro nutrients, Zn and Boron are the major yield limiting in pulses. However, the responses of pulse crops to these micronutrients differed widely with soil types and sensitivity of crops. Assessed by soil and plant analysis and confirmed through fertilizer response experiments, Zn deficiency plagues nearly 50 per cent of the Indian soils and crops. The data generated by AICRP on Micronutrients on the response of pulses to applied Zn and B indicate that the mean response to applied Zn varied from 160 to 460 kg/ha in different pulse crops.

9.5 Role of IPNS

Integrated plant nutrient supply (IPNS) as well as balanced fertilization is conceptually the same. The IPNS aims at maintenance on adjustment of soil fertility and of plant nutrient supply to an optimum level for sustaining the desired crop productivity through optimization of benefit from all possible sources of plant nutrients in an integrated manner. Balance fertilization must be based on the concept of integrated nutrient management for a cropping system as this is the only viable strategy advocating accelerated and enhance use of fertilizers with matching adoptions of organic manures and bio-fertilizers so that productivity is maintained for a sustainable agriculture. The balance has to be made in the soil crop system over time and has to take care of all other factors of production and make allowances for residual effects of post fertilizer applications, biological N fixation, etc. and to ensure that there is no toxicity/ deficiency of any element. The need to integration has been further felt after observing the effect of decontrol of phosphatic and potassic fertilizers. For maintenance of soil fertility on a sustainable basis in intensively cropped areas, greater emphasis has to be placed on residue management, inclusion of legumes as grains/fodder

9.5.1 Balanced Nutrition

To obtain the maximum benefit from every dollar spent, fertiliser programs must provide a balance of required nutrients. There is little

Table 9.3: Integrated Plant Nutrient Supply for Rice Based Cropping System in Different Agro-climatic Regions

Sl.No.	*Agro-climatic Regions*	*Cropping System*	*IPNS Recommendation*
1	Western Himalayan Region	Rice-wheat	**Rice**: 40 kg N + FYM/Green Manure @ 15 t/ha + 20 kg Zinc Sulphate (in Zn deficient soils)
			Wheat: 120 kg N + 80 kg P_2O_5 (through SSP) + 40 kg K_2O
2	Eastern Himalayan Region	Rice-wheat	**Rice**: 20 kg N + 20 kg P_2O_5 15 kg k_2O + FYM/GM@ 10 t/ha + Azolla @ 10 t/ha + 20 kg Zinc Sulphate once in 3 years + 5 kg Borax + 1 kg ammonium molybdate+5 kg copper sulphate
			Rice: 60 kg N+40 kg P_2O_5 + 25 kg K_2O + Azolla @ 10 t/ha
		Rice-wheat	**Rice**: 40 kg N+20 kg P_2O_5+40 kg K_2O+FYM@5 t/ha/GM + Azolla @10 t/ha+20 kg Zinc Sulphate once in 3 years + 5 kg borax + 1kg ammomium molybdate + 5 kg copper sulphate
			Wheat: 50 kg N + 20 kg P_2o_5 + FYM@5 t/ha
3.	Lower Gangetic Plain	Rice-wheat	**Rice**: 60 kg N + 40 kg P_2O_5 + 30 kg K_2O + FYM/GM @ 10 t/ha + 20 kg Zinc Sulphate
			Rice: 90 kg N+80 kg P_2O_5+60 kg K_2O + Azolla @ 10 kg/ha
		Rice-wheat	**Rice**: 40 kg N+45 kg P_2O_5+30 kg K_2O+FYM/GM@ 10 t/ha + Azolla @ 10 t/ha/BGA @ 10 kg/ha + 20 kg Zinc Sulphate
			Wheat: 90 kg N + 45 kg P_2O_5 + 45 kg K_2O
4.	Middle Gangetic Plain	Rice-wheat	**Rice**: 50 kg N + 30 kg P_2O_5 + 20 kg K_2O + green Manure (greengram/stover)+20kg Zinc Sulphate (in calcareous soils)
			Wheat: 90kg N+60kg P_2O_5+30 kg K_2O+FYM@10 t/ha OR
			Rice: 75 kg N + 45 kg P_2O_5 + 30 kg K_2O + BGA @ 15 kg/ha + FYM @ 10 t/ha + 20 kg Zinc Sulphate (in calcareous soils)
			Wheat: 100 kg N + 65 kg P_2O_5 + 30 kg K_2O

Contd...

Table 9.3–Contd...

Sl.No.	*Agro-climatic Regions*	*Cropping System*	*IPNS Recommendation*
5.	Upper Gangetic Plain	Rice-wheat	**Rice**: 90 kg N + 30 kg K_2O + FYM/GM (Sesbania/Leucaena lopping) @ 10 t/ha
			Wheat: 90 kg N + 60 kg P_2O_5 (through SSP) + 30 kg K_2O
6.	Trans Gangetic Plain	Rice/Cotton/ Maize/Bajra-wheat	**Rice**: 60 kg N + 30 kg K_2O + FYM/poultry manure/GM @ 10 t/ha
			Maize: 70 kg N + FYM/GM (sesbania/cowpea) @ 10 t/ha
			Cotton: 120 kg N
			Bajra: 60 kg N + 30 kg P_2O_5 + FYM @ 10 t/ha
			Wheat: 150 kg N + 30 kg P_2O_5 (through SSP) + 30 kg K_2O + Azotobacter/Azospirillium + PSB
7.	Eastern Plateau & Hills Plain	Rice-winter Maize/ Wheat/Pulses	**Rice**: 30 kg N + 15 kg P_2O_5 (through SSP) + 15 kg K_2O + FYM/GM @ 10 t/ha + 15 kg BGA
			Winter Maize: 100 kg N + 45 kg P_2O_5 (through SSP) + 20 kg K_2O
			Wheat: 90 kg N + 45 kg P_2O_5 (through SSP) + 30 kg K_2O
			Pulses: 10 kg N + 20 kg P_2O_5 (through SSP) + FYM @ 2.5 t/ha + Rhizobium + 500 g PSB
8.	Central Plateau & Hill Plain	Rice-Wheat/ Mustard	**Rice**: 75 kg N + FYM/Green Manure @ 5 t/ha
			Wheat: 90 kg N + 45 kg P_2O_5 + 30 kg K_2O
			Mustard: 30 kg N + 15 kg P_2O_t +10 kg K_2O FYM @ 10 t/ha
9.	Southern Plateau and Hills, East Coast Plains and Ghats and West Coast Plain Regions	Rice-Rice	**Rice**: 75 kg N + 15 kg P_2O_5 + 15 kg K_2O + FYM/green Manure @ 5 t/ha
			Rice: 90 kg N + 60 kg P_2O_5 + 40 kg K_2O + Azolla @ 10 t/ha/BGA @ 10 kg/ha + 20 kg Zinc Sulphate

point in applying enough N if P or Zn deficiency is limiting yield. To make better crop nutrition decisions, growers need to consider the use of paddock records, soil tests and test strips.

9.5.2 Nutrient Removal

The most important nutrient deficiencies in several areas are nitrogen (N), phosphorus (P), potassium (K), and zinc (Zn). Sulphur (S) and molybdenum (Mo), calcium (Ca) and magnesium (Mg), may be deficient on the more acid soils.

9.5.3 Fertiliser Application Techniques

Fertiliser can be applied before or at planting, side banded, as a foliar spray or through irrigation. Plants deficient in nitrogen (N) are stunted with pale green or yellow lower leaves, often with reddish tints and produce grain with low protein levels.

9.5.4 Calculation of Profile Nitrogen Levels from Soil Test Values

Most soil nitrogen test results are expressed in milligrams per kilogram (mg/kg) or parts per million (ppm). To make crop recommendations, it is necessary to convert nitrogen test results to kilograms of nitrogen per hectare (kg N/ha). The formula is:

$$\text{kg N/ha} = \frac{\text{Soil test value} \times \text{soil bulk density} \times \text{sample depth (cm)}}{10}$$

9.5.5 N Fertiliser Application

On the basis of paddock records or soil tests quantity that nitrogen fertiliser is required, can be used to obtain the quantity of actual fertiliser product required. For example, if 40 kg N/ha is required, this rate of nitrogen can be supplied by applying 87 kg/ha of urea. The formula to calculate the rate of nitrogen fertiliser product required to supply a required amount of nitrogen is:

$$\begin{array}{c}\text{Fertiliser Product}\\ \text{required (kg/ha)}\end{array} = \frac{\begin{array}{c}\text{Rate of nitrogen required}\\ \text{kg N/ha} \times 100\end{array}}{\begin{array}{c}\text{Per cent nitrogen in}\\ \text{fertiliser product}\end{array}}$$

9.6 A New Concept of Nutrient Management

Scientists, through research in major rice-growing areas across Asia, established the scientific basis for a new concept. The demand of a rice crop for essential nutrients is established by setting a yield target well-matched to local climatic and crop-growing conditions. The crop's need for supplemental nutrients is determined from the gap between the crop's demand for nutrients and the supply of these nutrients from existing native sources, including soil, crop residues, organic materials, and irrigation water.

9.6.1 The Right Practice at the Right Time

Scientists have found a way for rice farmers to increase their profit and produce more food by optimally applying essential nutrients to their crops. Rice requires essential nutrients such as nitrogen, phosphorus, and potassium that are typically not present in the soil in sufficient amounts to meet crop needs. The approach developed by scientists for optimal application of supplemental nutrients enables farmers to achieve rice yields well-matched to their local climatic and crop-growing conditions. With SSNM, farmers manage nutrients to match the needs of the crop for optimal growth at critical stages. "The need of a rice plant for nutrients varies, depending on its growth stage. When rice is young, its growth rate is slow, and thus, the need for nitrogen fertilizer is small. As rice grows older, its need for nitrogen increases," says Dr. Shaobing Peng, scientist at International Rice Research Institute, based in the Philippines [20].

9.6.2 Site-Specific Nutrient Management

Fertilizers are one of the main inputs in rice production. The quantity and management of fertilizers that best match the needs of rice crops for essential nutrients can vary greatly among fields, seasons, and years as a result of differences in crop-growing conditions, crop and soil management, and climate. Hence, the management of nutrients for rice requires an approach, which enables adjustments in applying N, P, and K to accommodate the field-specific needs of the rice crop for supplemental nutrients. Site-specific nutrient management (SSNM) in rice provides a field-specific approach for dynamically applying nutrients to crop as and when needed. This approach advocates optimal use of indigenous nutrients originating from soil, plant residues, manures, and irrigation water. Fertilizers are then applied in a timely fashion to overcome the deficit in nutrients between the total demand by crop to achieve a yield target and the supply from indigenous sources. Site-specific nutrient management (SSNM) is a simple plant need-based approach for optimally applying N, P, and K to different crops. It has been implanted in rice. The SSNM approach involves the following three steps.

Step 1: Establish an Attainable Yield Target

Crop yields are location and season specific depending upon climate, crop cultivars, and crop management. The yield target for a given location and season is the estimated grain yield attainable with the farmers' crop management when N, P, and K are effectively supplied. Because the amount of a nutrient taken up by a crop is directly related to yield, the yield target indicates the total amount of the nutrient that must be taken up by the crop. The yield target typically does not exceed about 80 per cent of climatic and genetic potential yield.

Step 2: Effectively Use Existing Nutrients

The SSNM approach promotes the optimal use of naturally occurring indigenous nutrients coming from the soil, organic amendments, crop residue, manure, and irrigation water. The uptake of a nutrient from indigenous sources can be estimated from the nutrient-limited yield, which is the grain yield for a crop not fertilized with the nutrient of interest but fertilized with other nutrients to ensure they do not limit yield.

Step 3: Apply Fertilizer to Fill the Deficit Between Crop Needs and Indigenous Supply

Fertilizer N, P, and K are applied to supplement the nutrients from indigenous sources and achieve the yield target. The quantity of required fertilizer is determined by the deficit between the crop's total needs for nutrients as determined by the yield target and the supply of these nutrients from indigenous sources as determined by the nutrient-limited yield (unfertilized plot). The required fertilizer N is distributed in several applications during the crop growing season to best match the crop's need for supplemental N. Fertilizer P and K are applied in sufficient amounts to overcome deficiencies and maintain soil fertility. It enables farmers to dynamically adjust fertilizer use to fill the deficit the nutrient needs of a high-yielding crop and the nutrient supply from between naturally occurring indigenous sources such as soil, crop residues, manures, irrigation water. The SSNM approach aims to apply nutrients at optimal rates and times to achieve high crop yields and high efficiency of nutrient use by the crop. It does not specifically aim to either reduce or increase fertilizer use. It is based on scientific principles developed through nearly a decade of on-farm research throughout Asia.

The initial concept of SSNM was developed in the mid-1990s and then evaluated from 1997 to 2000 in about 200 irrigated rice farms at eight sites in six Asian countries. The SSNM practices developed and evaluated in farmers' fields before 2001 increased yield and profit as compared to farmers' fertilizer practices. Since 2001 the initial SSNM concept had been systematically transformed through collaboration with national agriculture research and extension systems in Asia into a simplified framework for dynamic plant-need based management of N, P, and K for rice. Refined SSNM recommendations have been developed and evaluated through on-farm research involving thousands of farmers in major rice-growing areas of Bangladesh, China, India, Indonesia, Myanmar, the Philippines, Thailand, and Vietnam. The key features of the SSNM approach in diversified cropping include (1) dynamic adjustments in fertilizer N, P, and K management. It should be accommodate field- and season-specific according to crops grown in field; (2) Effective use of indigenous nutrients according to crops grown in field; (3) Fertilizer N management through the use of the leaf colour chart (LCC), which helps ensure N application with respect to time and amount needed by the crop; (4) use of the omission plot technique to determine the requirements for P and K fertilizer, and (5) managing fertilizer P and K to both overcome P and K deficiencies and avoid the mining of these nutrients from the soil. The total amount of required fertilizer N can be approximated from the anticipated crop response to fertilizer N application, which is the difference between attainable target yield and N-limited yield (*i.e.*, yield with no fertilizer N and no limitation of other nutrients).

9.7 Tools for Farmers

Among the tools developed for farmers is the leaf colour chart (LCC) used to estimate leaf nitrogen content. The LCC is a plastic ruler-shaped strip containing four or more panels ranging in colour from yellowish green to dark green. Farmers adjust their rates and timing of nitrogen fertilizer based on the colour of rice leaves. Dark green leaves indicate little or no immediate need for nitrogen. Yellowish green leaves indicate a relatively higher and urgent need of the crop for nitrogen fertilizer. Nitrogen is a mobile nutrient both in plant and soils and its efficiency in rice is only 30 to 40 per cent. Rest of the amount is lost through ammonia volatilization, denitrification and finally by nitrate leaching in underground water[21]. Generally, one-third to one-half of the fertilizer N applied at last puddling or before rice transplanting is lost. It is evident that rice seedlings take about 7 days to recover from transplanting shock and most of the basal applied N is not utilized efficiently. Leaf colour intensity is directly related to leaf chlorophyll content which in turn is related to leaf N status. LCC is a simple device that can be used by farmers for determining the right time of nitrogen application [22].

On the basis of Tables 9.4 & 9.5 it was concluded that the leaf colour chart appeared to be an easy and inexpensive tool for efficient N management in transplanted rice.

9.8 Limitation

9.8.1 Small Land Holdings

The average size of an operational holding is 1.57 ha small farm size has major implications for fertilizer management practices.

9.8.2 Poor Infrastructure Facilities

In India there are 519 soil testing laboratories. The total analyzing capacity of these laboratories is about 6.5 million samples per annum. In order to provide soil test-based fertilizer recommendations the existing analyzing capacity of the soil testing laboratories needs to be augmented almost 15-20 times.

9.8.3 Lack of Participatory Approach

Soil fertility will only be maintained and enhanced by the actions of farmers. Farmer's knowledge is essentially local, based on observation and experience within specific farming systems and agro-ecological contexts. Hence farmers' participation is important.

9.8.4 Low Availability of Organic Resources

The annual potential of organic resources ranges between 10.5-16.2 mt of NPK, only around 3.9-5.7 mt of plant nutrients can be made available for agricultural use. Average organic manure use at

Table 9.4: Schedule of Nitrogen Application Under Different Treatments During Experimentation [23]

Treatment	N Application Kg/ha at Different Days After Transplanting							Total
	Basal	18	26	32	39	46	62	
T_1 = Control	0	0	0	0	0	0	0	0
T_2 = Recom. N (120 Kg N/ha)	60	0	30	0	0	30	0	120
T_3 = N_{30} at LCC<3 no basal	0	30	0	0	30	0	0	60
T_4 = N_{30} at LCC<3 with basal	30	30	0	0	30	0	0	90
T_5 = N_{20} at LCC<5 no basal	0	20	0	20	20	20	20	100
T_6 = N_{20} at LCC<5 with basal	20	20	0	20	20	20	20	120
T_7 = N_{30} at LCC<5 no basal	0	30	0	30	30	30	30	150
T_8 = N_{30} at LCC<5 with basal	30	30	0	30	30	30	30	180

Table 9.5: Grain Yield, Total N Uptake, Agronomic Efficiency, Apparent Recovery of Nitrogen of Rice as Influenced by Different Nitrogen Management Practices at Srinagar (J&K)[23]

Treatment	*Grain Yield (t/ha)*		*Total N Uptake (kg/ha)*		*Agronomic Efficiency (kg grains/kg N)*		*Apparent Recovery of Nitrogen (per cent)*	
	2004	*2005*	*2004*	*2005*	*2004*	*2005*	*2004*	*2005*
T_1	3.03	2.76	68.8	64.1	–	–	–	–
T_2	4.86	4.81	107.2	108.4	15.3	17.0	32.0	36.9
T_3	4.42	4.66	95.2	94.0	23.2	31.6	44.0	49.8
T_4	4.57	4.82	98.8	99.7	17.2	22.8	33.3	39.5
T_5	5.96	6.03	120.9	121.1	29.4	32.7	52.1	57.0
T_6	6.03	6.15	125.9	126.4	25.0	28.2	47.6	51.9
T_7	6.46	6.45	135.0	134.4	22.9	24.6	44.1	46.9
T_8	6.61	6.70	138.2	138.9	20.0	21.9	46.2	42.1
SEm ±	0.25	0.25	6.1	8.6	1.2	1.6	1.6	1.9
CD (P=0.05)	0.74	0.73	18.4	25.4	3.6	4.9	4.9	5.9

present is about 2 tonnes ha^{-1}. The coverage under green manure crop is about 6 m ha and the use of bio-fertilizer, against a total bio-fertilizer demand of 1 mt, the current supply is less than 10,000 tonnes. Only 25 per cent nutrient needs of Indian agriculture can be met by utilizing various organic resources namely FYM (200 mt), crop residue (30 mt), urban/rural wastes (10 mt) and green manuring (25 m t).

9.8.5 High Labour Demand

For its mobilization, processing and application, because of low nutrient content and bulkiness it requires high labour.

9.9 Future Strategies

9.9.1 Awareness

Greater awareness needs to be created among the farmers for the use of farm resources on generation and its proper recycling and encouragement for the production of compost and green manuring.

9.9.2 Attention for Major Components

The major components of the system needs attention are: recycling of solid wastes and crop residues by composting and vermicomposting, more popularization of janata bio-gas plants, encouraging growth of legumes as part of the crop rotation for grain and fodder purposes, using sewage sludge and effluents for agriculture, integration of green manures, green leaf manures. BGA and azolla in rice culture.

9.9.3 Popularizing the Use of Biofertilizers

Popularizing bio-fertilizers to augment N and P supply by improving/strengthening transportation, distribution and storage infrastructure. Also enhancement of shelf life of Bio-fertilizers, development of new strains and easy technique for viability test for bio-fertilizers.

9.9.4 Soil Test Techniques

It needs to be refined so as to reduced the time, manpower and cost of chemicals during estimation and soil test laboratories should be strengthened and upgraded for soil and plant analysis, promoting balanced use of chemical fertilizers on soil testing and correction of secondary and micro nutrients deficiencies in soils.

9.9.5 Promoting Various Approaches

Advantages of introduction of green legumes in the cropping systems should be promoted. Use of phospho-compost should be promoted to supplement phosphatic fertilizer to a great extent. Research on incorporation N fixing ability in non-legumes need to be accelerated.

9.9.6 Maximize Nutrient Use

Promotion of appropriate soil, water and nutrient management and other agronomic practices to maximize nutrient use efficiently and economically for sustainable agricultural production and food security.

9.10 References

1. Gill, M.S., Pal, S.S., Ahlawat, I.P.S., 2008. Approaches for sustainability of rice (Oriza sativa)-wheat (*Triticum aestivum*) cropping system in Indo-Gangetic plains of India–A review. Indian Journal of Agronomy. 53 (2): 81-96.
2. Munda, G.C., Islam, M., Panda, B.B., 2008. Effect of organic and inorganics on productivity and uptake of nutrients in rice (*Oryza sativa*)-toria (*Brassica compestris*) cropping system. Indian Journal of Agronomy. 53 (2): 107-111.
3. Singh, Raj, Singh, B., Patidar, M., 2008. Effect of preceding crops and nutrient management on productivity of wheat (*Triticum aestivum)*-based cropping system in arid region. Indian Journal of Agronomy. 53 (4): 267-272.
4. Yadav, R.S., Yadav, P.C., Dhama A.K., 2003. Integrated nutrient management in wheat (*Triticum aestivum*)-mungbean (*Phaseolus radiatus*) cropping sequence in arid region. Indian Journal of Agronomy. 48 (1): 23-26.
5. Chand, S., Somani L.L., 2005. Exploring possibilities for improving the yield of mustard (*Brassic juncea* L. Czern. & Coss) through integrated nutrient management. International journal of tropical agriculture. 23: 177-182.
6. Singh, S.P., Dhyani, B.P., Shahi, U.P., Ashok, Kumar, Singh, R.R., Yogesh, Kumar, Sumit, Kumar, Balyan, V., 2009. Impact of integrated nutrient management on yield and nutrient uptake of rice (*Oryza sativa*) and wheat (*Triticum aestivum*) under rice-

wheat cropping system in sandy loam soil. Indian Journal of Agricultural Sciences. 79 (1): 65-69.

7. Tiwari, R.K., Khan, I.M., Singh, N., 2008. Effect of integrated nutrient management on transplanted rice (*Oryza sativa*) and its effect on succeeding wheat (*Triticum aestivum*). National Symposium on New Paradigms in Agronomic Research organized by Indian Society of Agronomy at Navsari Agricultural University, Navsari, Gujarat from November 19-21. p 18-19.

8. Virdia, H.M., Mehta, H.D., Bafna, A.M., Patel, Z.N., Gami, R.C., 2008. Integrated nutrient management in transplanted rice (*Oryza sativa*). National Symposium on New Paradigms in Agronomic Research organized by Indian Society of Agronomy at Navsari Agricultural University, Navsari, Gujarat from November 19-21. p 50-51.

9. Shivakumar, B.G., Ahlawat, I.P.S., 2008. Integrated nutrient management in soybean (*Glycine max*)-wheat (*Triticum aestivum*) cropping system. Indian Journal of Agronomy. 53 (4): 273-278.

10. Qureshi, F., Thomas, T., Bashir, U., 2009. Effect of integrated nutrient management on yield and nutrient availability for fieldpea Cv. Rachana under subtropical conditions. SKUAST J. Res., 11: 102-105.

11. Hegde, D.M. 1997. Nutrient requirement of solanaceous vegetable crops. Extension Bulletin-ASPAC Food Fert.Tech.Centre. 44: 9.

12. Singh, S.R., Sant, P., Kumar, J., 2000. Organic farming technology for sustainable vegetable production in H.P. Himachal Pradesh J. Agri. Res., 26: 69-73.

13. Harmson, K. (1995) Integrated phosphorus management. In integrated plant nutrition systems. FAO fertilizer and plant nutrition bulletin 12: pp 293-306.

14. Swarup, A.; Manna, M.C. and Singh, G.B. 1999. Impact of land use and management practices on organic carbon dynamics in soils of India. In: Global Climate Change and Tropical Ecosystems. (eds. R. Lal, J.M. Kimble, H. Eswaran & B.A. Stweart). Lewis Publishers, Boca Raton, Fl, pp. 261-281.

15. http://www.indiaagronet.com/indiaagronet/Technology_Upd/contents/integrated_nutrient_management_i.htm

16. Bandyopadhyay, S. and R. De. 1986. Ferti. Res. 10: 73-82.

17. Sharma, P. D. and Biswas, P.P. 2004. Fertilizer News Vol. 49(10). pp 43-47.

18. Prasad, R. Dinesh Kumar, S. N. Sharma, R.C. Gautam and M. K. Dwivedi (2004). Fertilizer News Vol. 49 (12). Dec. 2004, pp 73-80 (8 pages).

19. Hegde, D.M. and N. Sudhakar Babu (2004). Fertilizer News Vol. 49 (12) Dec. 2004, pp 103-110, 113-114 & 131 (11 pages).

20. http://www.irri.org/irrc/ssnmrice/

21. Prasad, R. 1999. Sustainable agriculture and fertilizer use. Current Science 77 (1): 38-43.

22. Balasubramanian, V., Ladha, J.K., Gupta R.K., Naresh,R.K., Mehla, R.S., Singh, B. and Singh, Y. 2003. Technology options for rice in the rice-wheat system in South Asia. (In) Improving the productivity and sustainability of rice-wheat system: Issues and Impact, pp 115-18. Ladha J.K. *et al.* (Eds) ASA Spec. Publ. 65. ASA, CSSA and SSSA, Madison,WI.

23. Singh, D.K., Singh, J.K. and Singh, L. 2009. Real time nitrogen management for higher N-use efficiency in transplanted rice (*Oryza sativa*) under temperate Kashmir conditions. Indian Journal of Agricultural Sciences 79 (10): 772-5.

Chapter 10

Principals and Practices of Organic Farming

10.1 Concept and Definition of Organic Farming

Globally, Organic Farming Systems have invited increasing attention for last three decades or so. They are perceived to offer some solutions to the problems currently besetting the agricultural sector of industrialized/green revolution countries.

As per Codex Committee on Food Labeling (FAO/WHO) "organic agriculture is a holistic production management system, which promotes and enhances agro-ecosystem health, including biodiversity, biological cycles, and soil biological activity. It emphasizes the use of management practices in preference to the use of off-farm inputs, taking into account that regional conditions require locally adapted systems. This is accomplished by using, wherever possible, agronomic, biological and mechanical methods, as opposed to using synthetic materials, to fulfill any specific function within the system". World Board of IFOAM (March 2008) has defined as "Organic agriculture is a production system that sustains the health of soils, ecosystems and people. It relies on ecological processes, biodiversity and cycles adapted to local conditions, rather than the use of inputs with adverse effects. Organic agriculture combines tradition, innovation and science to benefit the shared environment and promote fair relationships and a good quality of life for all involved". [1]

10.2 Why Organic Farming ?

- ☆ For balanced supply of nutrients (primary, secondary and micronutrients)
- ☆ To improved physical, chemical and biological properties of soil.
- ☆ It promotes low external input agriculture (LEIA)
- ☆ It restricted purchased inputs.
- ☆ It enhanced environmental security.
- ☆ To produced healthy and nutritionally superior food for man and animal.
- ☆ Organically grown plants are more resistant to diseases and pests and hence require less protective measures.

10.3 Benefits of Organic Farming

- ☆ Contributes in preservation of biodiversity.
- ☆ Produces healthy food.
- ☆ Improves health of soil
- ☆ Low water consumption
- ☆ Low input cost
- ☆ High produce cost
- ☆ High demand due to social awareness
- ☆ Huge export potential
- ☆ Promotion of sustainable agriculture for small farmers
- ☆ Ensures jobs in agriculture, food processing and marketing.

10.4 Principles of Organic Farming

Organic Agriculture is based on four basic principles and any system using the methods of organic agriculture and being based on these principles is organic agriculture [2]

- ☆ *The principle of health*: Organic Agriculture should sustain and enhance the health of soil, plant, animal, human and planet as one and indivisible.

- *The principle of ecology*: Organic Agriculture should be based on living ecological systems and cycles, work with them, emulate them and help sustain them.
- *The principle of fairnes*: Organic Agriculture should build on relationships that ensure fairness with regard to the common environment and life opportunities.
- *The principle of care*: Organic Agriculture should be managed in a precautionary and responsible manner to protect the health and well-being of current and future generations and the environment.

10.5 Aims of Organic Farming

- Resource conservation–Healthy Soil, Water, Biodiversity (plant/animal/microbes), Non-renewable energy
- Environmental Protection
- Sustainable productivity
- Healthy food–Nutritious, wholesome, free of pollutants
- Agribusiness

10.6 Organic Farming in India

- Area under certified organic farming≥2.55 m ha. (However, non-certified area is much more than certified area in the country).
- Total production–586 thousand tonnes.
- Total export of certified organic products–19.5 thousand tonnes–worth rupees 3012 million.
- No. of farmers certified under organic production–142 thousand.
- Accredited Inspection and certification agencies–12 (*e.g.* ECOCERT, Bureau Veritas Certification India Pvt. Ltd., IMO Control Pvt. Ltd., INDOCERT etc.).
- Organic standards under National Programme on Organic Production (NPOP), which are in harmony with EU standards.
- National Centre for Organic Farming established during X-plan to act as facilitator for promotion of organic farming

through: Capacity Building through Service providers. Setting up of vermiculture hatchery, Biofertiliser plant, Fruit/Vegetable compost plant (25 per cent Back ended bankable project). Human Resource Development through training, field demonstration. Setting up model organic farm. Quality testing and input production technology. Market development, publicity etc.

☆ Large number of states; Uttaranchal, Karnataka, Madhya Pradesh, Maharashtra, Gujrat, Rajasthan, Tamil Nadu, Kerala, Nagaland, Mizoram, Sikkim are promoting organic farming through different mechanisms.

10.7 Major Products of Organic Farming in India

Category	*Products*
Cereals	Rice, Wheat, Minor millets
Spices & Condiments	Cardamom, Black pepper, Ginger, Turmeric, Vanilla, Mustard, Tamarind, Clove, Cinnamon, Nutmeg, Mace, Chilly, Garlic, Onion, Cashewnut, Walnut
Pulses	Red gram, Black gram
Fruits	Mango, Banana, Pineapple, Grape, Passion fruit, Orange
Vegetables	Okra, Brinjal, Tomato, Potato
Oilseeds	Sesame, Castor, Sunflower
Others	Tea, Coffee, Cotton, Herbal extracts

10.8 Cultivation Techniques of Organic Farming[3]

10.8.1 Nutrient Management

Green Manures–*In situ* or *ex situ*

☆ Green manure crop supplies organic matter as well additional nitrogen particularly if it is a legume crop.

☆ A leguminous crop producing 8 to 25 tonnes of green matter per hectare will add about 60 to 90 kg of nitrogen when ploughed under.

☆ This amount would equal to application of 3 to 10 tonnes of farmyard manure on the basis of organic matter and its nitrogen contribution.

- Green manure crops are also helpful in protecting soil erosion and leaching of nutrients.
- They have soil ameliorating effect also.

Common Green Manure Crops

- *Sesbania aculeata* (Dhaincha)–Very popular among farmers. Best for saline/sodic soils. 45-60 days old crop.
- *Sesbania rostrata*–Nodules on stem also. Introduced from southeast Asia. Thrives well under flooded soils.
- *Crotalaria juncea* (Sunnhemp)–Multi-purpose (fodder/fibre/green manure).
- *Vigna Unguiculata* (Cow pea)–Multi-purpose (grain/fodder/vegetable/green manure)
- *Vigna radiata* (Green gram)–Used for grain/green manure

Common Green Leaf Manure Shrubs & Trees

- *Glyricidia maculata* (Glyricidia)
- *Leucenae leucocephala* (Subabool)
- *Robinia psudocasia* (Robinia)

Microbial Inoculants

Use of symbiotic/non-symbiotic microorganisms mainly as seed dressers and may also apply in field directly. Perform better under (nutrient) stress conditions.

Different Species of Bacteria Used as Biofertilizer

- Azotobacter (*A. chrococum*): Cereals, millets, vegetables, cotton, sugarcane etc.
- Beijerinckia (*B. indica*): Rice, sugarcane, forage grasses, coconut, areca nut, cashew, cocoa, pepper etc.
- Azospirillum (*A. lipoferum/A. brasiliense*): Rice, sorghum, sunflower, maize etc.
- Rhizobium (*R. japonicum/R. leguminoserum/R. phaseoli/trifolii* etc.): Legume crops
- Blue green algae (Cynobacteria)–*Nostoc, Anabaena, Aulosira, Calothrix* etc.
- Azolla (*A. pinnata–Anabaena association*): Low land rice

- PSM (Phosphate solubilizing microorganisms- Bacteria, fungi, actinomycete)–*Bacillus spp., Pseudomonas straita, Aspergillus awamori, A. acra, Penicillium digitam, Trichoderma spp.*–Solubilize low soluble P.
- VAM (Vesicular Arbuscular Mycorrhiza): Wheat, maize, millets, beans, potato, soybean etc.
- PGPRs (Plant Growth Promoting Rhizobia)–*Pseudomonas spp., Agrobacteria, Rhizobium spp., Bacillus spp.*: Enhance plant growth indirectly.

Composts

Organic wastes are converted into organic manures by means of biological activity under controlled conditions[4 and 5].

Different Kinds of Composts

- Based on method of composting–Indore, NADEP, Coimbatore, Banglore, Vermicompost etc.
- Based on material used–Animal-waste compost, leaf compost, Coir-pith compost etc.
- Enriched compost–Phospho-compost.
- Industrial-wastes/city-waste composts are not permitted.

Other Sources of Nutrient Supply

- Biogas Slurry
- Farmyard Manure
- Non-edible Oilcakes
- Crop Residues
- Legume Crops
- Rock Phosphate

10.8.2 Use of Mulches

We use mulches, because they:

- Conserve the soil moisture.
- Improve the infiltration of rain and irrigation water.
- Moderate soil thermal regimes.
- Protect the soil organisms, including earthworms.

- ☆ Suppress the weed growth.
- ☆ Release nutrients after decomposition and Improve soil organic matter content.

10.8.3 Use of Cover Crops

- ☆ Helpful in minimizing soil erosion both during summer and rainy season.
- ☆ Helpful in effective weed management.
- ☆ Helpful in moisture conservation.
- ☆ Helpful in improvement of soil biological complex.
- ☆ Helpful in maintenance and improvement of soil fertility and productivity.

10.8.4 Adaptation of Crop Rotation/Intercropping

Creation of bio-diversity in temporal and spatial dimensions

- ☆ Avoidance of exploitation of specific soil layer due to different rooting systems and nutrient needs.
- ☆ Management of soil-borne diseases/insect-pests and weeds with the help of Allelopathic effects, smothering effects, discontinuance of life cycle, distracting effects.
- ☆ Improving soil fertility by N-fixation.

10.8.5 Irrigation Practices in Organic Forming

- ☆ Enhance soil moisture retention and increase soil organic matter through application of mulches, preferably organic mulches. Need-based interculture operations and Adaptation of water harvesting techniques–*In situ/ex situ* methods.
- ☆ Use water saving methods like: Drip irrigation, Sprinkler irrigation, Raised beds planting.

10.8.6 Weed Management

A. Preventive Measures

- ☆ Prevent influx of new weed seeds–use weed free seed material, use only well decomposed manures.

- ☆ Prevent dissemination by eliminating them before seed dispersal.
- ☆ Soil cultivation methods–zero tillage, weed cures before sowing.
- ☆ Sowing time and plant density–optimal sowing time, closer spacing in row crops, enhanced plant densities.
- ☆ Crop rotations, Intercropping, Living green cover, Mulches etc.

B. Mechanical Measures

- ☆ Hand weeding.
- ☆ Tillage operations–Tilling, harrowing, ploughing, puddling.

C. Biological Measures

- ☆ Limited success.
- ☆ In an ecosystem there are at least 5-6 major weeds.
- ☆ No bio-control agent will be effective against all of them.
- ☆ Control of one weed will result in dominance of others.
- ☆ But successful in areas where a single weed dominates.
- ☆ Ideal for controlling weeds in non-crop areas, aquatic systems etc.

D. Cultural Methods

- ☆ Crop rotation
- ☆ Stale seed bed
- ☆ Optimum plant population
- ☆ Competitive crop cultivars
- ☆ Optimum planting date
- ☆ Optimum planting geometry
- ☆ Selective stimulation of crop
- ☆ Use of 'live mulches' or smother crops
- ☆ Intercropping

10.8.7 Plant Protection (Insect-pests and Disease Management)

Integration of All Methods

- Cultural (crop rotation, intercropping, sowing time, summer ploughing, insect/disease tolerant/resistant varieties etc) metods.
- Mechanical (light traps, pheromone traps etc) methods.
- Biological (parasites, predators, bio-pesticides etc.) methods.
- ITK-based methods (Biodynamic products, wood ash, vermi wash etc.)

Some Microbial Inoculants Used in Bio-control of Plant Pathogens and Crop Pests

- *Pseudomonas fluorescens*
- *Bacillus subtilis*
- *Bacillus thuringiensis*
- *Trichoderma harzianum*
- *Trichoderma viridae*
- *Bacillus thuringiensis*
- *Neumorea riley*
- *Metarhizum anisoplae*
- *Beauveria bassiana*
- *Verticillium lecanii*
- NPV

Some Other Important Examples of Bio-Agents/Bio-pesticides

- Neem-based products
- Garlic-chilli paste
- *Trichogramma japonicum*
- *Trichogramma chilonis*
- *Chrysoperla carnae*

10.9 Conclusions and Future Line of Work

- In the present modern agriculture, it is an urgent need to utilize all the locally available sources of agricultural inputs.
- Awareness about benefits of organic farming in farmers needs to be created through training, demonstration and visits.
- A computer database is needed for efficient utilization of natural sources.
- For restoring soil health *vis-à-vis* for quality products organic farming need to be popularize.

10.10 References

1. Organic Farming in India http://www.apeda.com, access on 17/01/2009.
2. Lampkin, N.H. (1994) Organic farming:Sustainable agriculture in practices- pp1-9. In: The economics of organic farming- and international prospective (Lampkin. N.H. and Padel, S. eds.). CAB International, U.K.
3. Prasad, K. and Gill, M.S. (2008) Organic farming–development and strategies. National symposium on new paradigsms in agronomic research organized by Indian society of agronomy at NAU, Navsari, Gujarat, pp. 501-502.
4. Chand, S. and Pabbi S. (2005) Organic farming–a rising concept. In Souvenir of agriculture summit-2005, organized by Ministry of Agriculture, Govt of India & FICCI, New Delhi, pp. 1-7.
5. Chand, S., Sahi, N.C. and Ali, T. (2006) Vermicomposting in organic farming In Souvenir of agriculture summit-2005, organized by Ministry of Agriculture, Govt of India & FICCI, New Delhi, pp. 1-4.

Appendices

Fertilizer Conversion Factors

In case if farmers want to use any other fertilizers against urea for nitrogen, DAP for phosphorus and MOP for potassium then they may convert the amount of available fertilizer against given amount of urea, DAP and MOP as per requirement of different crops

Appendix I. Nitrogenous Fertilizer

Sl.No.	*Fertilizer*	*N per cent*	*Equivalent to Urea (Factor)*
1.	Ammonium sulphate	20.6	2.233
2.	Ammonium chloride	25	1.84
3.	Ammonium nitrate	33.5	1.373
4.	Calcium ammonium nitrate	25	1.84
5.	Ammonium sulphat nitrate	26	1.77
6.	Urea	46	1
7.	DAP (for nitrogen)	18	2.556
8.	1 kg DAP = 0.391 kg urea for compensation of nitrogen if DAP not applied		

Appendix II. Phosphorus Fertilizer

Sl.No.	*Fertilizer*	*P_2O_5 per cent*	*Equivalent to DAP (Factor)*
1.	Single super phosphate (grad I)	16	2.875
2.	Single super phosphate (grad II)	14	3.286
3.	Triple super phosphate (grad I)	43	1.07
4.	Rock phosphate	30	1.53
5.	Phosphate	17	2.7
6.	Diammonium phosphate (DAP)	46	1

Appendix III: Potassium Fertilizer

Sl.No.	Fertilizer	K_2O per cent	Equivalent to MOP (factor)
1.	Potassium sulphate	48	1.25
2.	Potassium chloride (MOP)	60	1

Appendix IV: Nitrogen, Phosphorus and Potassium Content (per cent) in Different Organic Fertilizers

Sl.No.	Fertilizer	N per cent	P_2O_5 per cent	K_2O per cent
	Manures			
1.	FYM or cattle dung manure	0.5-1.5	0.4-0.8	0.5-1.9
2.	Urban compost	1.2-2.0	1.0	1.5
3.	Rural compost	0.4-0.8	0.3-0.6	0.7-1.0
4.	Green manure (Av. of different crops)	0.5-0.7	0.1-0.2	0.6-0.8
5.	Mushroom spent compost	1.2	0.7	0.85
6.	Poultry manure	2.87	2.9	2.35
7.	Sheep manure	1.5-3.0	1	2
8.	Vermicompost	1.5-2.6	0.9-1.7	1.5-2.4
9.	Sewage and sludge	1.5-3.5	0.8-4	0.3-0.6
	Cakes			
10.	Caster cake	5.5-5.8	1.8-1.9	1.0-1.1
11.	Mahua cake	2.5-2.6	0.1-0.9	1.8-1.9
12.	Karanj cake	3.9-4.0	0.9-1.0	1.3-1.4
13.	Neem cake	5.2-5.3	1.0-1.1	1.4-1.5
14.	Sunflower cake	4.8-4.9	1.4-1.5	1.2-1.3
15.	Coconut cake	3-3.2	1.8-1.9	1.7-1.8
16.	Groundnut cake	7.0-7.2	1.5-1.6	1.3-1.4
17.	Linseed cake	5.5-5.6	1.1-1.5	1.2-1.3
18.	Mustard cake	5.1-5.2	1.8-1.9	1.1-1.3
19.	Sesame cake	6.2-6.3	2.0-2.1	1.2-1.3

Contd...

Appendix IV–Contd...

Sl.No.	*Fertilizer*	*N* *per cent*	P_2O_5 *per cent*	K_2O *per cent*
	Others			
20.	Dry blood	10.0-12.0	1.0-1.5	0.6-0.8
21.	Fish manure	4.0-10.0	3.0-9.0	0.3-1.5
22.	Raw bone meal	3-4	20.0-25.0	-
23.	Steamed bone meal	1-2	25-30	-
24.	Night soil	1.2-1.3	0.8-1.0	0.4-0.5
25.	Human urine	1.0-1.2	0.1-0.2	0.2-0.3
26.	Cow urine	1.21	Trash-0.01	1.35
27.	Cow dung + cow urine mixture	0.6	0.15	0.45
28.	Crop residue	0.5	0.6	1.5
29.	Sheep and goat urine	1.47	0.05	1.96
30.	Rice husk	0.3-0.4	0.2-0.3	0.3-0.5
31.	Bagasse	0.25	0.12	-
32.	Press mud	1.25	2	-
33.	Tea waste	0.3-0.35	0.4	1.5
34.	Coir waste	1.26	0.06	1.2

Appendix V: Nitrogen, Phosphorus and Potassium Content (per cent) in Different Green Manuring Crops on Dry Weight Basis

Sl.No.	*Manure*	*N* *per cent*	P_2O_5 *per cent*	K_2O *per cent*
	Green manures			
1.	*Sesbania aculata*	2.34-3.3	0.24-0.7	1.3-2.97
2.	*Sesbania speciosa*	2.7	0.5	2.2
3.	*Crotolaria juncea*	2.21-2.6	0.38-0.6	1.478-2.0
4.	*Vigna catjang*	2.31	0.55	2.393
5.	*Phaseolus aurious*	2.2	0.48	2.103
6.	*Phaseolus trilobus*	2.1	0.5	2.1
7.	*Medicago sativa*	3.4	0.59	3.007
8.	*Tephrosia candida*	3.15	0.503	2.3
9.	*Tephrosia purpuria*	2.4	0.3	0.8
10.	*Glycin species*	2.84	0.614	1.359

Contd...

Appendix V–Contd...

Sl.No.	Manure	N per cent	P_2O_5 per cent	K_2O per cent
	Green leaf manure			
11.	*Pongamia glabra*	3.2	0.3	1.3
12.	*Gliricidia maculeata*	2.9	0.5	2.8
13.	*Azadorachta indica*	2.8	0.3	.4
14.	*Calatropis gigantea*	2.1	0.7	3.6

Appendix VI: Fertilizers may Also Contains Micronutrients in ppm

Sl.No.	Fertilizer	Copper	Zinc	Manganese	Boron	Molybdenum
1.	Ammonium sulphate	Trash-0.5	0.33	70	6.0	0.1
2.	Urea	0-3.6	0.5	0.5	0.5	0.7-6.2
3.	Calcium ammonium nitrate	Trash-18.0	8.35	10.50	Trash	-
4.	Single super phosphate	26.0	50-165	65-270	9.5	3.3
5.	Triple super phosphate	2-12	53-100	175-245	529	9.1
6.	Basic slage	9.2-56.4	4-59	68900	33.4	10.0
7.	Rock phosphate	5.6-9.5	24-137	130320	16	5.6
8.	Bone meal	270	660	500	715	-
9.	Potassium chloride	3.0	3.0	8.0	14.0	0.2
10.	Potassium sulphate	5.6-10.4	2.0	2.2-13.0	4.0	0.2
11.	Ammonium phosphate	3-4	80	115-220	-	2.2

Appendix VIIa: Chemicals and Minerals Used for Secondary Macro- and Micro-nutrient Fertilizers as Basal or Foliar Applications

VIIa: Calcium Fertilizer

Sl.No.	Fertilizer Name	Chemical Name	Calcium %
1.	Calcium ammonium nitrate		10-20
2.	Di calcium phosphate		32
3.	Single super phosphate		25-30
4.	Triple super phosphate		17-20
5.	Basic slage		33
6.	Lime		34

Table VIIb: Magnesium Fertilizer

Sl.No.	Fertilizer Name	Chemical Name	Mangnesium %
1.	Magnesite		40
2.	Mangnesium sulphate		16
3.	Multiplex (chileted)		10
4.	Dolomite		5-20

Table VIIc: Sulphur Fertilizer

Sl.No.	Fertilizer Name	Chemical Name	Sulphur %
1.	Ammonium sulphate	$(NH_4)_2SO_4$	24
2.	Ammonium sulphate nitrate	$(NH_4)_2NO_3.SO_4$	12
3.	Gypsum	$CaSO_4.2H_2O$	18.6
4.	Single super phosphate	$Ca(H_2PO_4)$ $H_2O.2CaSO_4$	10-12
5.	Potassium sulphate	K_2SO_4	17.6
6.	Potassium magnesium sulphate	$K_2(MgSO_4)_2$	22
7.	Magnesium sulphate	$MgSO_4$	17.6
8.	Manganese sulphate	$MnSO_4.5H_2O$	13.0
9.	Ferrus sulphate	$Fe(SO_4)_3$	12.8
10.	Copper sulphate	$CuSO_4.5H_2O$	18.8
11.	Zinc sulphate	$ZnSO_4.7H_2O$	17.8

Table VIId: Manganese Fertilizer

Sl.No.	Fertilizer Name	Chemical Name	Manganese %
1.	Manganese sulphate	$MnSO_4.3H_2O$	26-28
2.	Manganese oxide	MnO	41-68
3.	Manganese methoxicanil propen	Mn MPP	10-12
4.	Manganese chilet	Mn EDTA	12
5.	Manganese carbonet	$MnCO_3$	31
6.	Manganese chloride	$MnCl_2$	17
7.	Manganese oxide	MnO_2	63

Table VIIe: Iron Fertilizer

Sl.No.	Fertilizer Name	Chemical Name	Iron %
1.	Ferrus sulphate	$FeSO_{4.}\ 7H_2O$	19
2.	Ferric sulphate	$Fe_2(SO_4)_3.4H_2O$	23
3.	Ferrus oxide	FeO	77
4.	Ferric oxide	Fe_2O_3	69
5.	Ferrus ammonium phosphate	$Fe(NH_4)PO_4.H_2O$	29
6.	Ferrus ammonium sulphate	$(NH_4)_2SO_4.\ FeSO_4.6H_2O$	14

Table VIIf: Zinc Fertilizer

Sl.No.	Fertilizer Name	Chemical Name	Zinc %
1.	Zinc sulphate (mono hydrate)	$ZnSO_4.H_2O$	36
2.	Zinc sulphate (hepta hydrate)	$ZnSO_4.7H_2O$	22
3.	Zinc oxide	ZnO	60-80
4.	Zinc carbonate	$ZnCO_3$	56
5.	Zinc chloride	$ZnCl_2$	45-52
6.	Zinc phosphate	$Zn_3(PO_4)_2$	50
7.	Zinc nitrate (liquid)	$Zn(NO_3)_2$	15
8.	Stelrite	ZnS	60
9.	Zinc ammonium sulphate		10
10.	Zinc manganese ammonium sulphate		15
11.	Zinc dust		99.8

Table VII g: Copper Fertilizer

Sl.No.	Fertilizer Name	Chemical Name	Copper %
1.	Copper sulphate (mono hydrated)	$CuSO_4.H_2O$	35
2.	Copper sulphate (penta hydrated)	$CuSO_4.5H_2O$	25

Contd...

Table VII g–Contd...

Sl.No.	Fertilizer Name	Chemical Name	Copper %
3.	Basic copper sulphate	$CuSO_4.3Cu(OH)_2$	13-53
4.	Malakite	$CuCO_3.Cu(OH)_2$	57
5.	Asurite	$2CuCO_3.Cu(OH)_2$	55
6.	Cuprite	Cu_2O	89
7.	Cupric oxide	CuO	75
8.	Chalcosite	Cu_2S	80
9.	Chalcopyrite	$Cu2FeS_2$	35
10.	Copper acetate	$Cu(C_2H_3O_2)_2.H_2O$	32
11.	Copper ammonium phosphate	$Cu(NH_4)PO_4.H_2O$	32

Table VIIh: Boron Fertilizer

Sl.No.	Fertilizer Name	Chemical Name	Boron %
1.	Borex	$Na_2B_4O_7.10H_2O$	11
2.	Sodium penta borate	$Na_2B_{10}O_{16}.10H_2O$	18
3.	Sodium tetra borate (fertilizer borate-46)	$Na_2B_4O_7.5H_2O$	14
4.	Sodium tetra borate (fertilizer borate-65)	$Na_2B_4O_7$	20
5.	Solubor	$Na_2B_4O_7.5H_2O$	20
6.	Boric acid	$Na_2B_{10}O_{16}.10H_2O$	17
7.	Colmarite	H_3BO_3	10
8.	Boron frits		2-6
9.	Calcium borate	$Ca_2B_6O_{11}.5H_2O$	10

Table VIIi: Molybdenum Fertilizer

Sl.No.	Fertilizer Name	Chemical Name	Molybdenum %
1.	Sodium molybdate	$Na_2MoO_4.2H_2O$	39
2.	Ammonium molybdate	$(NH_4)_6Mo_7O_{24}.4H_2O$	54
3.	Molybdenum trioxide	MoO_3	66
4.	Molybdenum sulfide	MoS_2	60
5.	Molybdenum frits		2-3
6.	Molybdic oxide		47-66

Appendix VIII: Nitrogen Content and C/N Ratio of Some Compostable Materials

Materials	*Nitrogen Content (%)*	*C:N Ratio*
	Farm Residue	
Rice straw	0.3-0.5	80-130
Wheat straw	0.3-0.5	80-130
Barley straw	0.3-0.4	100-120
Maize stalks and leaves	0.8	50-60
Cotton stalks	0.6	70
Sugarcane trash	0.3-0.4	110-120
Lucern residue	2.55	19
Green weeds	2.45	13
Water hyacinth	2.38	17.6
Seaweeds	2.1	
Azolla	2.5	
Red clover	1.9	19
Ferns	1.5	25
Flax	1.1	44
Fallen leaves	0.5-1.0	40.8
Grass clippines	2.15	20
Sesbania sp.	2.83	17.9
Neem cake	6.05	4.5
	Animal Shed Waste	
Cow dung	2	
Buffalo dung		
Horsedung	2.4	
Poultry	5	
Sheep	3.75	
Pig	3.75	
	Human Habitation Waste	
Night soil	4.0-6.0	6–10
Urine	15-18	0.8
Digested sludge	5.0-6.0	6
Biogas (ex-cattle) slurry	2	20.4

Contd...

Appendix VIII–Contd...

Materials	Nitrogen Content (%)	C:N Ratio
	Vegetable Residue	
Potato tops	1.6	27
Amaranthus	3.6	11
Cabbage	3.6	12
Lettuce	3.7	
Onion	2.6	15
Pepper	2.6	15
Tomato	3.3	12
Carrot (whole)	1.6	27
Turnip top	2.3	
Fruit waste	1.5	
Tobacco	3	
	Forest	
Leaves	0.5-1.0	40-80
Raw sawdust	0.25	208
Rotted sawdust	0.3	128
Mango sawdust	0.3	132

Index